Annals of the ICRP

ICRP PUBLICATION 138

Ethical Foundations of the System of Radiological Protection

Editor-in-Chief
C.H. CLEMENT

Associate Editor
H. OGINO

Authors on behalf of ICRP
K-W. Cho, M-C. Cantone, C. Kurihara-Saio, B. Le Guen,
N. Martinez, D. Oughton, T. Schneider, R. Toohey, F. Zölzer

PUBLISHED FOR

The International Commission on Radiological Protection

by

Please cite this issue as 'ICRP, 2018. Ethical foundations of the system of radiological protection. ICRP Publication 138. Ann. ICRP 47(1).'

CONTENTS

ICRP Publication 138

Editorial

ETHICS OF RADIOLOGICAL PROTECTION: THE CHALLENGE AHEAD

The International Commission on Radiological Protection (ICRP) makes recommendations on how to behave wisely in relation to exposure of people and the environment to ionising radiation. Radiation is an ever-present part of the natural world, so the question of whether people and the environment should be exposed to radiation is moot. The question is what *action*, if any, we should take when faced with existing sources of radiation, or when considering introducing new ones.

How to respond to that fundamental question is the core business of ICRP.

An excellent place to begin is by trying to understand the effects of exposure to radiation. This is the realm of science.

Radiation epidemiology relies on statistical analyses of observed effects on large populations (of people, animals, and even plants) that have been exposed to radiation. For effects of radiation exposure on humans, the gold standard today is the work being done by the Radiation Effects Research Foundation with survivors of the atomic bombs dropped on Hiroshima and Nagasaki. In recent decades, this has been supplemented by studies of other large groups, such as workers in mines and nuclear facilities, medical patients, and people exposed to radon in homes, to name just a few.

Radiation biology takes a different approach, looking at how radiation exposure affects people, plants, and animals at the individual, tissue, cellular, and even sub-cellular level. This work is often done in laboratories with cell cultures or mice.

More recent efforts try to combine information from both approaches to best understand the relationship between exposure to radiation and resulting biological effects.

Scientific facts are essential to understanding, but, alone, are not enough to decide what to do. Ethical values are the other ingredient necessary for making recommendations on how to behave in light of our scientific knowledge. ICRP also relies on experience to help make recommendations practical.

The current publication is the first by ICRP dedicated to eliciting and describing the ethical foundations of the system of radiological protection. It gives us a common language to discuss ethics in this context. This includes identifying and describing four core ethical values that drive all ICRP recommendations, and several procedural values that aid practical implementation.

This common language will help professionals in radiological protection, ethics, and other fields to more deeply examine and refine the ethical basis of ICRP recommendations. It will also be helpful for a wider audience, by making the ethical basis of radiological protection more transparent.

To this end, there is an intention to describe the ethical basis of ICRP recommendations more explicitly in future publications, just as we often describe the scientific basis in detail, using the language of the current publication as a foundation. Sometimes, this will be relatively straightforward, deserving of only brief mention. Where this publication will help the most is in more ethically complex situations. One example of this is the protection of animals not normally considered part of the environment. We are just beginning to explore this area, which includes, e.g., veterinary patients and animals used in scientific experiments.

We are also considering a companion to the current publication that will focus on the ethics of radiological protection in medicine. The strong body of research in biomedical ethics had a considerable influence on the development of this publication. In addition, medicine is by far the largest source of intentional exposure to radiation, making it a major part of ICRP's programme of work. Together, these factors argue that this is a logical next step, and a good opportunity to describe the core ethical values in a less abstract context.

So, while this publication is a major step forward in describing one of the three pillars of the system of radiological protection (scientific knowledge, ethical values, and practical experience), it is much more a beginning than an end. There is no doubt that the core ethical values need to be examined in more depth. There is also a need to balance these values, all of which are essential, and none of which are absolute. There is no a-priori hierarchy among them; a definite context is needed to use these values in deciding how best to act in a particular circumstance. This is the challenge that lies ahead.

CHRISTOPHER CLEMENT
ICRP SCIENTIFIC SECRETARY
EDITOR-IN-CHIEF

ICRP Publication 138

ETHICAL FOUNDATIONS OF THE SYSTEM OF RADIOLOGICAL PROTECTION

ICRP PUBLICATION 138

Approved by the Commission in October 2017

Abstract–Despite a longstanding recognition that radiological protection is not only a matter of science, but also ethics, ICRP publications have rarely addressed the ethical foundations of the system of radiological protection explicitly. The purpose of this publication is to describe how the Commission has relied on ethical values, either intentionally or indirectly, in developing the system of radiological protection with the objective of presenting a coherent view of how ethics is part of this system. In so doing, it helps to clarify the inherent value judgements made in achieving the aim of the radiological protection system as underlined by the Commission in *Publication 103*. Although primarily addressed to the radiological protection community, this publication is also intended to address authorities, operators, workers, medical professionals, patients, the public, and its representatives (e.g. NGOs) acting in the interest of the protection of people and the environment. This publication provides the key steps concerning the scientific, ethical, and practical evolutions of the system of radiological protection since the first ICRP publication in 1928. It then describes the four core ethical values underpinning the present system: beneficence/ non-maleficence, prudence, justice, and dignity. It also discusses how these core ethical values relate to the principles of radiological protection, namely justification, optimisation, and limitation. The publication finally addresses key procedural values that are required for the practical implementation of the system, focusing on accountability, transparency, and inclusiveness. The Commission sees this publication as a founding document to be elaborated further in different situations and circumstances.

Keywords: Radiological protection system; Ethical values; Procedural values

AUTHORS ON BEHALF OF ICRP
K-W. CHO, M-C. CANTONE, C. KURIHARA-SAIO, B. LE GUEN,
N. MARTINEZ, D. OUGHTON, T. SCHNEIDER, R. TOOHEY, F. ZÖLZER

PREFACE

The ICRP Main Commission established Task Group 94 of Committee 4 during its meeting in Fukushima, Japan in October 2012 to develop an ICRP publication presenting the ethical foundations of the system of radiological protection. To achieve this goal, the Commission asked the task group to review the publications of the Commission in order to identify the ethical values associated with the system of radiological protection for occupational, public, and medical exposures, and for the protection of the environment. In proposing this approach, the Commission recognises that the system of radiological protection has, during its evolution, been informed by ethical values.

Given the nature of the work, the Commission also encouraged Task Group 94 to develop its work in close cooperation with specialists of ethics and radiological protection professionals from around the world. With this in mind, a series of workshops was organised by ICRP in collaboration with the International Radiation Protection Association (IRPA) and academic institutions to fully examine, discuss, and debate the ethical basis of the current system of radiological protection. These workshops were held in: Daejeon (Korea) and Milan (Italy) in 2013; Baltimore (MD, USA) in 2014; and Madrid (Spain), Cambridge (MA, USA), and Fukushima (Japan) in 2015. Presentations were given to spur discussion in group sessions. Presenters were from a variety of backgrounds and fields of expertise.

Task Group 94 also benefited from discussions at an International Symposium on Ethics of Environmental Health in Budweis, Czech Republic in 2014; the Fourth Asian and Oceanic Congress on Radiation Protection in Kuala Lumpur, Malaysia in 2014; the UK Workshop on the Ethical Dimensions of the Radiological Protection System in London, UK in 2014; the Third International Symposium on the System of Radiological Protection in Seoul, Korea in 2015; and the 14th IRPA Congress in Cape Town, South Africa in 2016.

The membership of Task Group 94 was as follows:

K-W. Cho (Chair)	N.E. Martinez	R. Toohey
M-C. Cantone	D. Oughton	S. Wambani
S.O. Hansson	T. Schneider	F. Zölzer
C. Kurihara-Saio		

The corresponding members were:

R. Czarwinski	B. Le Guen	E. Van Deventer

The Committee 4 critical reviewers were:

F. Bochud   J. Takala

The Main Commission critical reviewers were:

C.M. Larsson   E. Vañó

Task Group 94 worked mainly by correspondence and met three times on 2 and 3 February 2015 at the Technical University of Madrid, Spain; and on 8–10 July 2015 and 26–28 January 2016 at Nuclear Protection Evaluation Centre, France. The task group wishes to thank the organisations and staff that made facilities and support available for these meetings.

In drafting the publication, Task Group 94 received significant contributions from ICRP Vice-Chair Jacques Lochard and ICRP Scientific Secretary Christopher Clement, and also received input from several participants of the workshops organised in cooperation with IRPA and the other organisations mentioned above.

The membership of Committee 4 during the period of preparation of this publication was:

(2009–2013)

J. Lochard (Chair)	T. Homma	A. McGarry
W. Weiss (Vice-Chair)	M. Kai	K. Mrabit
J-F. Lecomte (Secretary)	H. Liu	S. Shinkarev
P. Burns	S. Liu	J. Simmonds
P. Carboneras	S. Magnusson	A. Tsela
D.A. Cool	G. Massera	W. Zeller

(2013–2017)

D.A. Cool (Chair)	M. Doruff	A. Nisbet
K-W. Cho (Vice-Chair)	E. Gallego	D. Oughton
J-F. Lecomte (Secretary)	T. Homma	T. Pather
F. Bochud	M. Kai	S. Shinkarev
M. Boyd	S. Liu	J. Takala
A. Canoba	A. McGarry	

MAIN POINTS

- Radiological protection relies on scientific knowledge, ethical considerations, and practical experience. This is the first ICRP publication dedicated to elaborating the ethical foundations of the system of radiological protection.
- This publication provides a foundation and common language for discussion of the ethical aspects of radiological protection between experts and with a wider audience, and insight into applying the system of radiological protection where there are competing ethical priorities.
- The system of radiological protection relies on four core ethical values:
  - Beneficence/non-maleficence: promoting or doing good, and avoiding doing harm. This is reflected, for example, in the primary aim of the system of radiological protection:...an appropriate level of protection...without unduly limiting...desirable human actions.
  - Prudence: making informed and carefully considered choices without full knowledge of the scope and consequences of an action. Prudence is reflected, for example, in the consideration of uncertainty of radiation risks for both humans and the environment.
  - Justice: fairness in the distribution of advantages and disadvantages. Justice is a key value underlying, for example, individual dose restrictions that aim to prevent any individual from receiving an unfair burden of risk.
  - Dignity: the unconditional respect that every person deserves, irrespective of personal attributes or circumstances. Personal autonomy is a corollary of human dignity. This underlies, for example, the importance placed on stakeholder participation and the empowerment of individuals to make their own informed decisions.
- The core ethical values support the aims of the system of radiological protection and its three fundamental principles: justification, optimisation, and individual dose limitation.
- Three procedural values are highlighted to aid the practical implementation of radiological protection: accountability, transparency, and inclusiveness (stakeholder participation).

GLOSSARY

Accountability

The obligation of individuals or organisations who are in charge of decision making to answer for their actions to all those who are likely to be affected, including reporting on their activities, accepting responsibility, and accounting for actions taken and the consequences, if necessary.

Autonomy

The capacity of individuals to act freely, to decide for themselves, and to pursue a course of action in their life.

Beneficence

To promote or do good. Beneficence is a key value of biomedical ethics. In radiological protection, it is to increase the direct and indirect benefits for individuals, communities, and the environment.

Consequentialism (also called teleological ethics)

An approach to ethics that judges the morality of an action based on the action's impact on the well-being of people and the common good. Utilitarianism ethics is the most well-known variant of consequentialism.

Deontological ethics

An approach to ethics that judges the morality of an action based on the action's adherence to rules or duties.

Dignity

The value and respect that every person has and deserves regardless of her/his age, sex, health, social condition, ethnic origin, religion, etc.

Equity

The quality of being fair and impartial. In radiological protection, equity refers to the fair distribution of risks and benefits of radiation exposures.

Ethics

The branch of philosophy that explores the nature of moral virtue and evaluates human actions using sets of moral principles and concepts to govern behaviour or the conducting of an activity.

Fairness

The quality of treating people equitably and in a way that is reasonable.

Harm

Damage or injury that is caused by a person or an event.

Inclusiveness

Ensuring that all those concerned are given the opportunity to participate in discussions, deliberations, and decision making concerning situations that affect them.

Informed consent

The voluntary agreement to an activity based on sufficient information and understanding of the purpose, benefits, and risks.

Justice

The upholding of what is right, equitable, and fair.

- Distributive justice: fairness in the distribution of advantages and disadvantages among members of communities.
- Environmental justice: equitable distribution of environmental risks and benefits; fair and meaningful participation in environmental decision making; and recognition of community ways of life, local knowledge, and cultural differences.
- Intergenerational justice: fairness towards everyone, with attention also to future generations.
- Procedural justice: fairness in the rules and procedures in the process of decision making.
- Restorative justice: giving priority to repairing the harm done to victims, communities, and the environment.
- Social justice: promoting a just society by recognition of human rights to equitable treatment and assuring equal access to opportunities.

Non-maleficence

To avoid doing harm. Non-maleficence is a key value of biomedical ethics. In radiological protection, it is to reduce the direct and indirect harm and risk for individuals, communities, and the environment.

Practical radiological protection culture

The knowledge and skills enabling citizens to make well-informed choices and behave wisely in situations involving potential or actual exposure to ionising radiation.

Precautionary principle

A principle in risk management whereby measures are put in place to prevent or reduce risks when science and technical knowledge are not able to provide certainty.

Procedural values

Set of values to take practical actions that align the conduct of a given activity with the ethical principles.

Prudence

To make informed and carefully considered choices without the full knowledge of the scope and consequences of an action.

Radiation risk

The potential harm posed by exposure to radiation. In risk assessment, risk is a combination of the probability of occurrence of damage or injury and its severity.

Reasonableness

To make rational, informed, and impartial decisions that respect other views, goals, and conflicting interests.

Right to know

The right of individuals to be informed about what hazards they are exposed to and how to protect themselves.

Self-help protection

> Informed actions taken by individuals to protect themselves, their family, and their communities.

Stakeholder participation

> The participation of all relevant parties in the decision-making processes related to radiological protection. Also referred to as 'stakeholder involvement' or 'stakeholder engagement'.

Tolerability

> The degree or extent to which something can be endured.

Transparency

> Accessibility of information about the deliberations and decisions concerning potential or on-going activities, and the honesty with which this information is transmitted.

Utilitarian ethics

> An approach to ethics that judges the morality of an action based on the action's impact on individual and social welfare.

Value judgement

> A subjective assessment based upon available knowledge and a particular set of values and priorities.

Virtue ethics

> An approach to ethics that emphasises the role of personal character and virtue in determination of morality.

Wisdom

> The quality of having knowledge, common sense, experience, and good judgement in order to make reasonable decisions and to act accordingly.

1. INTRODUCTION

1.1. Background

(1) In an address to the Ninth Annual Conference on Electrical Techniques in Medicine and Biology in 1956, Lauriston Taylor, then incumbent President of the National Council on Radiation Protection and Measurements, and Chairman of the International Commission on Radiological Protection (ICRP), declared, 'Radiation protection is not only a matter for science. It is a problem of philosophy, and morality, and the utmost wisdom' (Taylor, 1957). By using the term 'wisdom', one of the fundamental virtues of many cultures worldwide, Taylor emphasised that beyond its undeniable and compelling scientific and ethical basis, radiological protection was also a question of insight, common sense, good judgement, and experience. Through his formulation, he brought to light three pillars of the system of radiological protection that have been built up gradually for almost half a century, namely science, ethics, and experience.

(2) Despite a longstanding recognition that radiological protection is not only a matter of science, but also relies on ethical considerations, ICRP publications have rarely addressed the ethical foundations of the system of radiological protection explicitly. This does not mean that the Commission has been unaware of the importance of such considerations. Protection recommendations inevitably represent an ethical position, irrespective of whether that position is explicit or implied. Therefore, ethical considerations can be traced in many ICRP publications.

(3) Regarding the ethical dimension of radiological protection, it should be pointed out at the outset that there are very few writings devoted to it compared with the vast literature related to the scientific, technical, and practical aspects. The first contributions directly addressing the subject only appeared in the 1990s. Among them, it is worthwhile mentioning the pioneering contribution of Giovanni Silini, who reviewed the ethical foundation of the system during the Sievert Lecture he delivered in 1992 (Silini, 1992). He concluded his lecture by emphasising that the system has been developed rationally, but at the same time with the desire to act reasonably. Also interesting to note are the articles published subsequently by academics questioning the ethical theories underpinning the system (Oughton, 1996; Schrader-Frechette and Persson, 1997) which ultimately led to the recognition that the system of radiological protection can be seen as being based on the three major theories of philosophical ethics that combine the respect of individual rights (deontological ethics), the furthering of collective interest (utilitarian ethics), and the promotion of discernment and wisdom (virtue ethics) (Hansson, 2007). In turn, inspired by these reflections, eminent professionals of radiological protection have seized the subject (Lindell, 2001; Clarke, 2003; Streffer et al., 2004; Clarke and Valentin, 2009; Gonzalez, 2011; Valentin, 2013; Lochard, 2016; Clement and Lochard, 2017). Worthwhile to note is also the cross-cultural approach to the ethics of radiological protection, exploring the commonalities between Western theoretical and applied ethics and written and oral traditions worldwide (Zölzer, 2013, 2016).

(4) This relatively recent interest in ethical aspects of radiological protection is certainly not unrelated to the difficulties encountered for decades by radiological protection professionals facing the questions and concerns of people. The traditional emphasis on the science of radiation by the Commission has been shown to be insufficient, and it is now acknowledged that human and ethical dimensions of exposure situations are also important, and sometimes decisive, in both the decision-making process and in communication, particularly when engaging with stakeholders.

(5) The lessons learned from the management of the consequences from the Chernobyl accident have certainly played a key role in this awareness (Oughton and Howard, 2012; Lochard, 2013), as have challenges from radioactive waste management (NEA/OECD, 1995; Streffer et al., 2011), increasing use of medical applications (Malone, 2013), and, more recently, the Fukushima accident. It is in this context that ICRP initiated a reflection on the ethical foundations of the system of radiological protection in early 2010 and a task group in 2012. In order to involve a wide variety of expertise in this process, the Commission invited ethicists, philosophers, social scientists, and radiological protection professionals from different regions of the world to a series of regional workshops organised in collaboration with the International Radiation Protection Association (IRPA) and academic institutions.

1.2. Scope and objective

(6) This publication reviews the Commission's previous publications to identify the ethical values associated with the ICRP system of radiological protection for occupational, public, and medical exposures, and for protection of the environment. It describes key components of the ethical theories and principles prevailing in the fields of health and the environment relevant to radiological protection.

(7) This publication aims to emphasise how the Commission has relied on ethical values in developing the system of radiological protection, with the objective of presenting a coherent view of how ethics is part of this system. Ethics cannot provide conclusive solutions, but can help to facilitate discussions among those seeking to promote the well-being of individuals, the sustainable development of society, and the protection of the environment. A clearer understanding of the core ethical values and related principles of radiological protection will help to address issues emerging from potential conflicts in decision making.

(8) A particular objective of this publication is to outline what can reasonably be expected from radiological protection to individuals and societies. In so doing, it helps to clarify the inherent value judgements made in achieving the aim of the system of radiological protection as underlined by the Commission in *Publication 103* (ICRP, 2007a), and thus hopefully facilitates decision-making processes and communication on radiation risk.

(9) Although primarily addressed to the radiological protection community, this publication is also intended for authorities, operators, workers, medical

professionals, patients, the public, and its representatives acting in the interest of the protection of people and the environment.

(10) The Commission recently adopted a Code of Ethics (ICRP, 2015b) setting out what is expected from its members in the development of its recommendations and guidance. This code emphasises the need for ICRP members to be committed to public benefit, and to act independently while being impartial, transparent, and accountable. Various professional societies have also developed codes of ethics for their members (e.g. IRPA, 2004). These behavioural requirements are beyond the scope of this publication, and are not discussed further here. However, the ethical values discussed in this publication can help to guide radiological protection professionals in the conduct of their duties.

(11) The work leading to this publication is the first concerted effort by the Commission to reflect upon and describe the ethical basis of the system of radiological protection in some detail. The Commission sees this publication as a founding document for addressing ethical issues in future recommendations. As such, it does not discuss in detail the questions and dilemmas that still exist in radiological protection, nor provide advice to any specific scenario involving radiation exposure. Initiating a discussion of both the ethical values and their implementation should make ethical reasoning more accessible to those working in the field, and hopefully encourage them to apply it explicitly in decisions and practices (Martinez and Wueste, 2016).

1.3. Structure of this publication

(12) Section 2 presents the milestones which marked the evolution of the system of radiological protection since the first ICRP publication in 1928. Section 3 describes the core ethical values that shape the system, and also discusses how these core ethical values underpin the principles of radiological protection, namely justification, optimisation, and limitation. Section 4 discusses key procedural values underlying the requirements for the practical implementation of the system. Section 5 summarises the main implications of ethics for the system of radiological protection. Annexes A, B, and C address ethical theories, biomedical ethical principles, and cross-cultural values relevant to radiological protection, respectively. Annex D provides lists of participants at the workshops on the ethics of the system of radiological protection.

2. EVOLUTION OF THE SYSTEM OF RADIOLOGICAL PROTECTION

(13) The present system of radiological protection is based on three pillars: the science of radiological protection, combining knowledge from different disciplines; a set of ethical values; and the experience accumulated from the day-to-day practice of radiological protection professionals. This is illustrated in Fig. 2.1. Explicit guidelines for balanced consideration of these three pillars in decision making are not often seen, seemingly because there is no direct, quantifiable way to do so: each pillar informs the others, yet has an individual nature that does not lend itself to a straightforward intercomparison. Moreover, each exposure situation has unique characteristics or circumstances that need to be considered in making a decision. As such, instead of a fixed, universal response, value judgements are required to assess a particular situation or circumstance, and determine how the pillars should be combined and applied in that instance.

(14) The present system has evolved with the advancement of science, the evolution of societal values, and the lessons of experience, and has matured to more clearly reflect the necessity of value judgements in interpreting risk and making appropriate decisions: 'All of those concerned with radiological protection have to make value judgements about the relative importance of different kinds of risk and about the balancing of risks and benefits' (ICRP, 2007a). The guiding actions for radiological

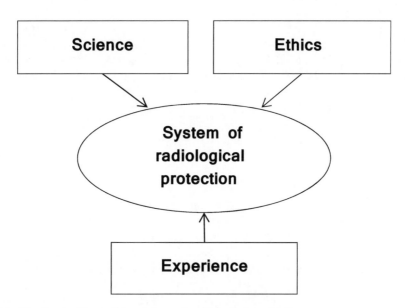

Fig. 2.1. The three pillars of the system of radiological protection.

protection have been governed by the following questions, which call for value judgements in their response:

- Are the circumstances generating exposure justified?
- Are all exposures maintained as low as reasonably achievable under the prevailing circumstances?
- Are the radiation doses received by individuals considered tolerable?

(15) To make value judgements, there must be corresponding knowledge about the circumstance and the possible implications of actions (information about what 'is'), and ethical values on which to base decisions for action (a sense of what 'should be'). As this publication is concerned with the ethical basis of the system of radiological protection, the focus here is on the pillar of core ethical values, with the intention of providing support for making value judgements. The following subsections describe how the system has evolved progressively in relation to the development of scientific knowledge of radiation effects, and the historical events associated with the use of radiation and radioactivity. Through these considerations, one can gain insight into the consistent set of core ethical values that has underpinned the present system since the beginning.

2.1. The early days: do no harm

(16) The first international recommendations on radiological protection were issued in 1928 by the International X-ray and Radium Protection Committee (IXRPC) (IXRPC, 1928), although some advice had been published much earlier (Fuchs, 1896). Three decades had passed since the discovery of x rays (Roentgen, 1895), natural radioactivity (Becquerel, 1896), and radium (Curie, 1898), during which time the use of radiation in medicine had increased significantly.

(17) The formation of the IXRPC (renamed ICRP in 1950) at the Second International Congress of Radiology, and its first recommendations, were prompted by the international medical community's desire to address the (sometimes serious) skin reactions being observed in some medical practitioners and investigators. The 1928 Recommendations focused squarely on protection of 'x-ray and radium workers' in medical facilities, and provided advice meant to avoid harmful skin reactions and derangements of internal organs and changes in the blood: 'the dangers of over-exposure...can be avoided by the provision of adequate protection'.

(18) This advice was based on the best scientific knowledge at the time about the effects of radiation exposure, the experience of three decades of practice, and the desire to avoid harm. The relatively simple, implicit ethical principle of 'doing no harm' was sufficient, as it was thought that straightforward protection measures could keep exposures low enough to avoid injury entirely. The only type of effects known at that time were deterministic effects, which are considered to have a threshold below which no deleterious effects are seen, although they were not described in these terms until decades later.

(19) Over the next two decades, the use of radiation continued to increase, not only in the medical field but also in the radium industry. To keep pace, the scope of the system expanded from protection of medical professionals to include radium workers. There was also an increasing understanding of the thresholds for various health effects. In the 1934 Recommendations (IXRPC, 1934), the concept of a 'tolerance dose' of 0.2 roentgens per day was introduced. Scientific advancements resulted in refinements in the measures to be taken to avoid doing harm, but the basic ethical principle of doing no harm remained.

(20) The 1950 Recommendations (ICRP, 1951) saw the first hints of the evolution of the ethical basis of the system beyond avoidance of doing harm, or at least that the practicalities of achieving this aim might be less straightforward than previously thought, recommending that 'every effort be made to reduce exposures to all types of ionising radiation to the lowest possible level'.

2.2. A more complex problem: managing risk, a matter of balance

(21) The 1950s saw a growing societal concern about the effects of exposure to radiation, not only to workers but also to the public and patients. This was fuelled by the atomic bombings of Hiroshima and Nagasaki in 1945 and the aftermath: the nuclear weapons testing after World War II causing global contamination, as well as highly publicised events such as the serious contamination of the population of the Marshall Islands, and the Japanese tuna fishing boat The Lucky Dragon No. 5, exposed to fallout from a US atomic bomb test in 1954 (Lapp, 1958).

(22) This growing concern, along with the increasing use of radiation in many fields including the nuclear energy industry, potential hereditary effects suggested by animal experiments, and emerging evidence of increased leukaemia in radiologists and atomic bomb survivors, had a profound influence on the system. The 1954 Recommendations (ICRP, 1955) stated that 'no radiation level higher than the natural background can be regarded as absolutely "safe"' and recommended that 'exposure to radiation be kept at the lowest practicable level in all cases'. Furthermore, it was in these recommendations that the system first incorporated protection of the public.

(23) Cancer and hereditary effects (also referred to as 'stochastic effects'), for which it was now assumed there is no absolutely safe level of exposure (no threshold), presented a much more ethically complex situation than before. It was no longer enough to avoid doing harm by keeping exposures sufficiently low. The main problem shifted from avoiding harm to managing the probability of harm.

(24) It took many years to develop the framework to deal with this complex situation. In *Publication 9* (ICRP, 1966), noting the absence of evidence as to the existence of a threshold for some effects, and in view of the uncertainty concerning the nature of the dose–effect relationship in the induction of cancers, the Commission saw '...no practical alternative, for the purposes of radiological protection, to assuming a linear relationship between dose and effect, and that doses act

cumulatively'. By adopting this position, the Commission was fully aware 'that the assumptions of no threshold and of complete additivity of all doses may be incorrect', but it considered that there was no alternative given the information available at that time (ICRP, 1966). Consequently, as any level of exposure to radiation was assumed to involve some degree of potential harm, the Commission added the objective of limiting the probability of occurrence of damage associated with stochastic effects.

(25) This was further elaborated in *Publication 26* (ICRP, 1977), where the primary aim of the system was described as 'protection of individuals, their progeny, and mankind as a whole while still allowing necessary activities from which radiation exposure might result'. As a consequence, protection was constrained to avoid interfering with 'necessary activities'. This publication also introduced the three basic principles of radiological protection (justification of practice, optimisation of protection, and limitation of individual doses), and was the first attempt to introduce considerations about tolerability of risk to derive individual dose restrictions. In *Publication 60* (ICRP, 1991), the primary aim of the system was reformulated to focus more on balancing the potentially competing priorities of the benefits of protection from radiation and the benefits of the use of radiation, rather than on constraining protection: 'to provide an appropriate standard of protection for man without unduly limiting the beneficial practices giving rise to radiation exposure'.

2.3. A broader perspective: protecting the environment

(26) More recently, the system also expanded from human to non-human species. *Publication 26* (ICRP, 1977) was the first to mention protection of the environment. However, it did not go beyond the assertion that 'if man is adequately protected then other living things are also likely to be sufficiently protected'. This statement, reworded, was repeated in *Publication 60* (ICRP, 1991): 'the standards of environmental control needed to protect man to the degree currently thought desirable will ensure that other species are not put at risk'.

(27) Over the next two decades, there was a broad increase in environmental awareness, and a rise in societal expectations that protection of the environment must be assured rather than assumed. These ideas took hold globally following the 1992 Rio Declaration on Environment and Development (UNCED, 1992). In parallel, extensive scientific work on the impact of radiation on non-human biota was undertaken by several organisations, which led protection of the environment against radiation to be treated more substantially in *Publication 91* (ICRP, 2003). The latter introduced the ICRP framework for assessing the impact of ionising radiation on non-human species.

(28) The elaboration of the framework included an explicit reflection on ethical values, touching on the different philosophical world views regarding how the environment is valued (i.e. anthropocentric, biocentric, and ecocentric approaches), and presenting a selection of internationally agreed principles concerning environmental

protection. These were sustainable development, conservation, preservation, maintenance of biological diversity, environmental justice, and human dignity. The publication also addressed procedural principles and operational strategies, including, amongst others, the precautionary principle, informed consent, and stakeholder participation.

2.4. Considering the diversity of exposure situations

(29) In recent decades, the system has been challenged by the widespread impact of the Chernobyl accident in 1986, the concern of malevolent acts following an increase in terrorist attacks during the last decade, as well as the increasing awareness of the legacy of areas contaminated by past activities and of the exposure associated with natural sources of radiation. More recently, the Fukushima Daiichi accident in 2011 has challenged the system again in much the same way.

(30) No doubt, the core of the system remains the protection of patients, workers, members of the public, and the environment from radiation sources introduced deliberately in the medical, industrial, and nuclear domains. Fortunately, these circumstances are usually well controlled. However, other exposure situations are more difficult to control, leading to complex societal issues arising from the associated exposures. As such, *Publication 103* (ICRP, 2007a) introduced the distinction between 'existing exposure situations', 'emergency exposure situations', and 'planned exposure situations' to take account of the degree of controllability of sources, exposure pathways, and the exposures of people.

(31) This new framework better recognises the distinct natures and associated challenges of exposure situations resulting from natural and man-made sources that exist before the decisions to control them are taken (e.g. cosmic radiation or legacy sites), as well as resulting from loss of control or intentional misuse of sources. A critical aspect of these complex situations is that the public may be faced with significantly higher exposure levels compared with those prevailing with planned exposure situations. Experience shows that affected people have to be directly involved to manage these situations effectively and fairly.

(32) In 1999, the importance of the participation of relevant stakeholders in making decisions about protection was recognised. However, it was not until *Publication 103* in 2007 that it was explicitly introduced in the general recommendation as 'the need to account for the views and concerns of stakeholders when optimising protection' (ICRP, 2007a). This recommendation was illustrated shortly thereafter in *Publication 111* (ICRP, 2009) with the introduction of self-help protection. This was to recognise the important role of stakeholder participation in the management of postaccident situations for individuals to make informed decisions in order to improve the radiological situation for themselves, their family, and their community. Such an approach implies a certain level of autonomy of individuals, relying on information, advice, and support from authorities and radiological protection experts.

2.5. The system of radiological protection today

(33) Today, the primary aim of the system remains 'to contribute to an appropriate level of protection for people and the environment against the detrimental effects of radiation exposure without unduly limiting the desirable human actions that may be associated with such exposure' (ICRP, 2007a). For human health, the system aims to 'manage and control exposures to ionising radiation so that deterministic effects are prevented, and the risks of stochastic effects are reduced to the extent reasonably achievable'. Put another way, effects that can be prevented are prevented, and effects for which the risk cannot be reduced to zero are managed through optimisation of protection, together with the application of dose restrictions. The current aim for protection of the environment is to avoid having anything more than a 'negligible impact on the maintenance of biological diversity, the conservation of species, or the health and status of natural habitats, communities and ecosystems' (ICRP, 2008).

(34) Serving these aims, the present radiological protection system encompasses three fundamental principles to achieve its objectives:

- The principle of justification, which states that any decision that alters the exposure situation should do more good than harm. This means that by introducing a new radiation source in planned exposure situations, or by reducing exposures in existing and emergency exposure situations, one should achieve sufficient benefit to offset any costs or negative consequences. The benefits are deemed to apply to specific individuals, society as a whole, and also to the environment.
- The principle of optimisation, which stipulates that all exposures should be kept as low as reasonably achievable, taking into account economic and societal factors. It is a source-related process, aimed at achieving the best level of protection under the prevailing circumstances through an ongoing, iterative process. This principle is the cornerstone of the system of protection. Furthermore, in order to avoid inequitable distributions of individual exposures, the Commission recommends restricting doses to individuals and non-human biota from a particular source.
- The principle of limitation, which declares that individual exposures should not exceed the dose limits recommended by the Commission. It applies only to planned exposure situations, other than medical exposure of patients or exposure of non-human biota.

(35) These three fundamental principles of protection are central to the system of radiological protection, which applies to different types of exposure situations (planned, emergency, and existing) and categories of exposure (occupational, public, medical exposure of patients, and environmental).

3. CORE ETHICAL VALUES UNDERPINNING THE SYSTEM OF RADIOLOGICAL PROTECTION

(36) As described in Section 2, although values were not referred to explicitly in ICRP publications during the development of the principles of justification, optimisation, and limitation, they played a key role throughout. The review of past publications of the Commission in the light of theoretical and applied ethics (see Annexes A and B) has led to the identification of four core ethical values underpinning the current system of radiological protection: beneficence/non-maleficence, prudence, justice, and dignity. These values, which are shared across cultures worldwide (see Annex C), are presented and discussed in the following subsections.

3.1. Beneficence and non-maleficence

(37) Beneficence means promoting or doing good, and non-maleficence means avoiding causation of harm (Frankena, 1963). These two related ethical values have a long history in moral philosophy, dating back to the Hippocratic Oath, which demands that a physician do good and/or not harm (Moody, 2011). They were formalised in modern biomedical ethics in the late 1970s following the publication of the so-called 'Belmont Report' (DHEW, 1979) and the related seminal work of philosophers Tom Beauchamp and Jim Childress (Beauchamp and Childress, 1979). The Commission has not previously used the terms 'beneficence' and 'non-maleficence', but they are central to the system of radiological protection.

(38) In its most general meaning, beneficence includes non-maleficence (Ross, 1930). Beneficence and non-maleficence can also be seen as two separate values. This publication treats them as a single value. By developing recommendations seeking to protect people against the harmful effects of radiation, the Commission undoubtedly contributes to serving the best interest of individuals and, indirectly, the quality of social life. This is achieved in practice by ensuring that deterministic effects are avoided and stochastic effects are reduced as far as achievable given the prevailing circumstances. Non-maleficence is closely related to prevention, which aims to limit risk by eliminating or reducing the likelihood of hazards, and thus promote well-being.

(39) In a narrower sense, beneficence includes consideration of direct benefits for individuals, communities, and the environment. The deliberate use of radiation, although coupled with certain risks, can undoubtedly have desirable consequences, such as the improvement of diagnostics or therapy in medicine, or the production of electricity. These have to be weighed against the potential harmful consequences. Similar considerations also apply to existing and emergency exposure situations.

(40) A key challenge for beneficence and non-maleficence is how to measure the benefits, harms, and risks. In radiological protection, this involves consideration of both their individual and societal aspects. From the viewpoint of evidence-based medicine and public health, a more comparative analysis of medical factors that

27

affect health is needed, including not only radiation but also other exposures. In addition, a variety of social, psychological, and cultural aspects need to be considered, and there may be disagreement on what matters, or on how to value or weight these factors. Nevertheless, it is recommended that such an assessment be transparent about what was included, recognise disagreements where they arise, and go beyond a simple balancing of direct health impacts against economic costs. In this respect, it is worth recalling the WHO definition of health: 'Health is a state of complete physical, mental and social well-being and not merely the absence of disease or infirmity' (WHO, 1948). As discussed in Section 4, participation of stakeholders other than radiological protection experts is a key part of such a holistic assessment.

(41) An evaluation of beneficence and non-maleficence must also address the question of who or what counts in evaluation of potential harms and benefits, including, for example, future generations and the environment. As mentioned previously, protection of the environment is now included in the primary aim of the system in *Publication 103* (ICRP, 2007a). One could ask whether environmental harm is being avoided for the sake of people (an anthropocentric view), or whether the environment is being protected for its own sake (a non-anthropocentric approach) (ICRP, 2003). ICRP does not endorse any specific approach, and considers both to be compatible with the value of beneficence and non-maleficence. In *Publication 124* (ICRP, 2014a), it is recommended that the evaluation of actual and potential consequences of human activities involving radiation should include, and integrate, effects on both humans and the environment, ensuring that the overall outcome results in more good than harm.

3.2. Prudence

(42) Prudence is the ability to make informed and carefully considered choices without full knowledge of the scope and consequences of actions. It is also the ability to choose and act on what is in our power to do and not to do.

(43) Prudence has a long history in ethics. It is considered to be one of the main virtues rooted in the Western tradition developed by Plato and Aristotle; the teaching of Confucius; the Hindu and Buddhist philosophies; and the ancient traditions of the peoples of Eurasia, Oceania, and America. Originally, prudence signifies 'practical wisdom', which is the meaning of the Greek word 'phronesis'. It describes the quality of having knowledge, experience, and good judgement to take reasonable decisions and to act accordingly.

(44) The system of radiological protection is based on solid scientific evidence; however, there are remaining uncertainties at low levels of exposure that necessitate value judgements. Decision making requires prudence as a central value. However, prudence should not be taken to be synonymous with conservatism or never taking risks. It describes the way in which decisions are made, and not solely the outcome of those decisions.

(45) It is worth noting that prudence appeared in the late 1950s (ICRP, 1959) in the Commission's recommendations in relation to the uncertainties related to stochastic effects. Since then, it has been constantly re-affirmed in relation to the linear no-threshold (LNT) model. Thus, in *Publication 103,* one can read: 'The LNT model is not universally accepted as biological truth, but rather, because we do not actually know what level of risk is associated with very-low-dose exposure, it is considered to be a prudent judgement for public policy aimed at avoiding unnecessary risk from exposure' (ICRP, 2007a).

(46) More specifically, the term 'prudence' is used explicitly in connection with the different types of effects of radiation exposure considered in the system.

- Deterministic effects: 'It is prudent to take uncertainties in the current estimates of thresholds for deterministic effects into account...Consequently, annual doses rising towards 100 mSv will almost always justify the introduction of protective actions' (ICRP, 2007a).
- Stochastic effects in general: 'At radiation doses below around 100 mSv in a year, the increase in the incidence of stochastic effects is assumed by the Commission to occur with a small probability and in proportion to the increase in radiation dose...The Commission considers that the LNT model remains a prudent basis for radiological protection at low doses and low dose rate' (ICRP, 2007a).
- For heritable effects in particular: 'There continues to be no direct evidence that exposure of parents to radiation leads to excess heritable disease in offspring. However, the Commission judges that there is compelling evidence that radiation causes heritable effects in experimental animals. Therefore, the Commission prudently continues to include the risk of heritable effects in its system of radiological protection' (ICRP, 2007a).

(47) Policy makers do not generally refer to prudence. Instead, reference is made to the precautionary principle, which was popularised by the Rio Conference on Environment and Development (UNCED, 1992). This principle, which states that lack of scientific certainty shall not be used to justify postponing appropriate measures 'where there are threats of serious or irreversible damage', has been much debated in connection with the ethics of decision making in recent years. This is also at stake in the domain of radiological protection (Streffer et al., 2004).

(48) Neither prudence nor the precautionary principle should be interpreted as demanding zero risk, choosing the least risky option, or requiring action just for the sake of action. The experience of over half a century of radiological risk management applying the optimisation principle can be considered as a reasoned and pragmatic application of prudence and/or the precautionary principle. Interestingly, the Commission mentions in its most recent recommendations that the use of the LNT model remains a prudent basis for radiological protection at low doses and low dose rates considered 'to be the best practical approach to managing risk from radiation exposure and commensurate with the "precautionary principle"' (UNESCO, 2005; ICRP, 2007a).

(49) The implications of this prudent attitude have been significant for the subsequent structuring of the system of radiological protection. A careful study of the evolution of the Commission's recommendations over the past decades shows that this central assumption led to gradually shaping the system as it stands now (Lochard and Schieber, 2000). This is clearly summarised by the Commission as follows: 'The major policy implication of the LNT model is that some finite risk, however small, must be assumed and a level of protection established based on what is deemed acceptable. This leads to the Commission's system of protection with its three fundamental principles of protection' (ICRP, 2007a).

(50) In addition, the adoption of a prudent attitude induces the duty of vigilance vis-à-vis the effects of radiation, resulting in an obligation to monitor radiological conditions for humans and non-human biota. Specifically, prudence requires that research should be open to unexpected findings, beyond the obligation to conduct relevant research in an attempt to reduce existing uncertainties (e.g. epidemiology, radiobiology, metrology, radio-ecology). Furthermore, for humans, prudence implies support of the exposed population, including – if necessary – detection and treatment of possible pathologies induced by ionising radiation.

3.3. Justice

(51) Justice is usually defined as fairness in the distribution of advantages and disadvantages among groups of people (distributive justice), fairness in compensation for losses (restorative justice), and fairness in the rules and procedures in the processes of decision making (procedural justice). Whereas equity and inequity relate to the state of affairs in distribution of goods, fairness can be used to describe the degree of equity attained in this distribution.

(52) It must be emphasised that the Commission has not referred to justice explicitly in its previous recommendations. However, the idea of limiting individual exposures in order to correct possible disparities in the distribution of individual doses due to radiation among exposed populations appeared as early as *Publication 26* (ICRP, 1977). In *Publication 60*, the term 'inequity' was used for the first time: 'When the benefits and detriments do not have the same distribution through the population, there is bound to be some inequity. Serious inequity can be avoided by the attention paid to the protection of individuals' (ICRP, 1991).

(53) Any exposure situation, whether natural or man-made, can result in a wide distribution of individual exposures. In addition, the implementation of protective measures can also induce potential distortions in this distribution that may aggravate inequities. In this context, the protection criteria of the system of radiological protection play a dual role.

(54) First, radiological protection criteria aim to reduce inequities in the distribution of individual exposures in situations where some individuals could be subject to much more exposure than others. This restriction of individual exposures is done through the use of dose constraints that apply to planned exposure situations,

reference levels that apply to existing and emergency exposure situations, and derived consideration reference levels that apply for the protection of fauna and flora. Dose constraints, reference levels, and derived consideration reference levels are integral parts of the optimisation process, and thus must be chosen depending on the prevailing circumstances by those responsible for protection.

(55) The second role of protection criteria is to ensure that exposures do not exceed the values beyond which the associated risk is considered as not tolerable given a particular context. This is ensured through the application of dose limits recommended by the Commission for the protection of workers and the public in planned exposure situations. As with dose constraints and reference levels, dose limits are tools to restrict individual exposure in order to ensure fairness in the distribution of risks across the exposed group of individuals. However, given the predictable dimension of the planned exposure situations for which the radiation sources are introduced deliberately by human action, the numerical values of dose limits, unlike dose constraints and reference levels, are generally specified in legal terms and have a binding character.

(56) Thus, through the protection criteria, the system of radiological protection aims to ensure that the distribution of individual exposures meets two principles of distributive justice. First, the principle of equity reflects the personal circumstances in which individuals are involved. It is the role of dose constraints and reference levels to reduce the range of exposure to individuals subject to the same exposure situation. Secondly, the principle of equal rights guarantees equal treatment for all individuals belonging to the same category of exposure in planned exposure situations. It is the role of dose limits to ensure that all members of the public, and all occupationally exposed workers, do not exceed the level of risk deemed tolerable by society and recognised in law (Hansson, 2007).

(57) Recognition of the right of citizens to participate in decision-making processes is an important aspect of procedural justice, and linked to stakeholder participation. In environmental justice, this has been ratified in the Århus Convention on Access to Information, Public Participation in Decision-making, and Access to Justice in Environmental Matters (UNECE, 2001). There are, of course, still challenges in achieving this in practice, and stakeholder participation is discussed in more detail in Section 4.

(58) Intergenerational distributive justice has been addressed by the Commission for the management of radioactive waste with reference to 'precautionary principle and sustainable development in order to preserve the health and environment of future generations' (ICRP, 2013, Para. 15). In *Publication 81*, the Commission recommends that 'individuals and populations in the future should be afforded at least the same level of protection as the current generation' (ICRP, 1998, Para. 40). In *Publication 122*, the Commission introduces responsibilities towards future generations in terms of providing the means to deal with their protection: '... the obligations of the present generation towards the future generation are complex, involving, for instance, not only issues of safety and protection but also transfer of knowledge and resources' (ICRP, 2013, Para. 17).

3.4. Dignity

(59) Dignity is an attribute of the human condition: the idea that something is due to a person because she/he is human. This means that every individual deserves unconditional respect, irrespective of personal attributes or circumstances such as age, sex, health, disability, social condition, ethnic origin, religion, etc. This idea has a prominent place in the Universal Declaration of Human Rights, which states that 'All human beings are born free and equal in dignity and rights' (United Nations, 1948). Dignity has a long history as the central value in many ethical theories, including Kant's notion to treat individuals as subjects, not objects: 'Act in such a way that you treat humanity, whether in your own person or in the person of any other, never merely as a means to an end, but always at the same time as an end' (Kant, 1785). Personal autonomy is a corollary of human dignity. This is the idea that individuals have the capacity to act freely (i.e. to make uncoerced and informed decisions).

(60) Respect for human dignity was first promoted in radiological protection as 'informed consent' in biomedical research, which means that a person has 'the right to accept the risk voluntarily' and 'an equal right to refuse to accept' (ICRP, 1992). Together with the concept of 'right to know', 'informed consent' was clearly established in *Publication 84* on pregnancy and medical radiation (ICRP, 2000). Beyond the medical field, human dignity was introduced explicitly as 'the need for the respect of individual human rights and for the consequent range of human views' in the elaboration of the ICRP framework for the protection of the environment (ICRP, 2003). The Commission has also emphasised the promotion of autonomy through stakeholder participation (ICRP, 2007a) and empowerment of individuals to make informed decisions, whether, for example, confronted with contaminated land (ICRP, 2009), security screening in airports (ICRP, 2014b), radon in their homes (ICRP, 2014c), or cosmic radiation in aviation (ICRP, 2016). The system of radiological protection thus actively respects dignity and promotion of the autonomy of people facing radioactivity in their daily lives. It is worth noting that the promotion of dignity is also related to a set of procedural ethical values (accountability, transparency, and stakeholder participation), developed in Section 4, which are linked to the practical implementation of the system of radiological protection.

3.5. Relationship between the core ethical values and the fundamental principles

(61) The four core ethical values permeate the current system of radiological protection, but their relationship with the three principles of justification, optimisation, and limitation is not straightforward. This is not so much the case for justification, which can be understood as mainly, although not exclusively, referring to beneficence/non-maleficence, or rather the balancing of 'doing good' and 'avoiding harm'. When it comes to optimisation (i.e. to keep exposure as low as reasonably

achievable, taking into account economic and societal factors) and dose limitation (i.e. to maintain risk at a tolerable level), these principles depend upon several of the core ethical values.

(62) The two key concepts of reasonableness and tolerability, which are central to the second and third principles, respectively, specify how the radiation risk is managed by combining and balancing the core ethical values (Schneider et al., 2016).

(63) The concept of reasonableness can be traced back to the 1950s when the Commission recommended that 'it is highly desirable to keep the exposure of large populations at as low a level as practicable' (ICRP, 1959). This recommendation evolved into the Commission's introduction of the optimisation principle two decades later (ICRP, 1977). There was first an attempt to define reasonableness using a quantitative approach, such as cost–benefit analysis (ICRP, 1983). Later, the search for reasonableness gradually led to the recognition that quantification alone was insufficient to reflect justice, both as fairness in the distribution of individual doses and as consideration for the concerns and views of stakeholders.

(64) The concept of tolerability is present from the early publications of the Commission (ICRP, 1959). In *Publication 60*, a conceptual framework was introduced which allows one to determine the degree of tolerability of an exposure (or of the associated radiation risk), and thus, depending on the category of exposure (public or occupational), to distinguish between unacceptable and tolerable levels of exposure (ICRP, 1991). In *Publication 103*, tolerability is referred to specifically in each type of exposure situation, taking into account not only the radiation risk associated with exposure (and the related value of non-maleficence), but also the practicality of reducing or preventing the exposure (prudence and beneficence), the benefits from the exposure situation to individuals and society (beneficence and justice), and other societal criteria (justice and dignity) (ICRP, 2007a).

(65) Applying the principles of radiological protection is a permanent quest for decisions that rely on the core ethical values underlying the system of radiological protection; in other words, to do more good than harm, avoid unnecessary risk, establish a fair distribution of exposures, and treat people with respect (Lochard, 2016). In this pursuit, the two concepts of reasonableness and tolerability, although supported by quantitative methods, definitively remain of a deliberative nature.

4. PROCEDURAL VALUES

(66) For the practical implementation of its recommendations, the Commission sets out a number of requirements relating to the procedural and organisational aspects of radiological protection. It does not go into detail, but merely lays down some broad standards, leaving the task of developing them to other international organisations (IAEA, 2014). Three of these requirements deserve to be highlighted because they are common to all exposure situations: accountability, transparency, and inclusiveness (stakeholder participation). All three have strong ethical aspects which will be considered in this section. It is also important to recognise that these procedural values are inter-related.

4.1. Accountability

(67) Accountability can be defined as the procedural ethical value that people who are in charge of decision making must answer for their actions to all those who are likely to be affected by these actions. In terms of governance, this means the obligation of individuals or organisations to report on their activities, to bear responsibility, and to be ready to account for the consequences if necessary. The concept of accountability appeared explicitly in *Publication 60* (ICRP, 1991), and was reaffirmed in much the same terms in *Publication 103* (ICRP, 2007a) addressing the implementation of the recommendations and in considering organisational features: 'In all organisations, the responsibilities and the associated authority are delegated to an extent depending on the complexity of the duties involved. (...) There should be a clear line of accountability running right to the top of each organisation. (...) Advisory and regulatory authorities should be held accountable for the advice they give and any requirements they impose'.

(68) The Commission also considered the accountability of the present generation to future generations, which is mentioned explicitly in *Publications 77* (ICRP, 1997b), *81* (ICRP, 1998), *91* (ICRP, 2003), and *122* (ICRP, 2013) related to waste management and the protection of the environment. As an example, *Publication 122* (Para. 17) states '...the obligations of the present generation towards the future generation are complex, involving, for instance, not only issues of safety and protection but also transfer of knowledge and resources. Due to the technical and scientific uncertainties, and the evolution of society in the long term, it is generally acknowledged that the present generation is not able to ensure that societal action will be taken in the future, but needs to provide the means for future generations to cope with these issues' (ICRP, 2013). Accountability in this context is part of implementing the value of intergenerational distributive justice discussed in Section 3.

4.2. Transparency

(69) Transparency is also part of implementing the value of procedural justice. It concerns the fairness of the process through which information is shared intentionally between individuals and/or organisations. According to the International Standards Organisation (ISO), transparency means 'openness about decisions and activities that affect society, the economy and the environment, and willingness to communicate these in a clear, accurate, timely, honest and complete manner' (ISO, 2010). Transparency does not simply mean communication or consultation. It relates to the accessibility of information about the activities, deliberations, and decisions at stake, and also the clarity, practicality, and honesty with which this information is transmitted. With respect to the accessibility of information, it is part of government and corporate social responsibility to ensure that decision makers act responsibly in the social, economic, and environmental domains in the interest of individuals and groups concerned in all exposure situations. Clearly, security or economic reasons can be put forward to justify the control or limitation of outgoing information from a business or an organisation. This is why explicit procedures must be in place, and expectations made clear, from the outset to allow for good transparency (Oughton, 2008).

(70) Transparency on exposures and protective actions for the workers has been integrated into ICRP recommendations since the 1960s. One can thus read: 'Workers should be suitably informed of the radiation hazard entailed by their work and of the precautions to be taken' (ICRP, 1966). This requisite has since been expanded in subsequent recommendations (ICRP, 1991, 2007a). It was not, however, until the 2000s that transparency became a general principle applicable not only to information about exposures, but also on the decision-making processes concerning the choices of protective actions. Moreover, it was generalised to all categories of exposure: occupational, medical, public, and environmental. This was introduced for the first time in *Publication 101b* dedicated to the optimisation of protection and bearing the evocative subtitle 'Broadening the process': 'Due to its judgemental nature, there is a strong need for transparency of the optimisation process. All the data, parameters, assumptions, and values that enter into the process must be presented and defined very clearly. This transparency assumes that all relevant information is provided to the involved parties, and that the traceability of the decision-making process is documented properly, aiming for an informed decision' (ICRP, 2006).

(71) In practice, transparency depends on the category of exposure and the type of exposure situation. In the medical field, it is implemented according to different modalities and procedures based on categories, such as through training for workers (ICRP, 1997a) and informed consent of patients (ICRP, 1992, 2007b). It also appears as the right to know principle for the public in the case of security screening, for example (ICRP, 2014b). In its latest recommendations, the Commission emphasised that '. . . scientific estimations and value judgements should be made clear whenever possible, so as to increase the transparency, and thus the understanding, of how decisions have been reached' (ICRP, 2007a). This shows that the requisite of transparency should apply wherever value judgements are involved in the system of radiological protection.

(72) Informed consent has been well developed in the context of biomedical ethics (e.g. biomedical research, radiotherapy, or interventional radiology), but is also important outside of the medical field. Prerequisite elements of informed consent include information (which should be appropriate and sufficient), comprehension, and voluntariness (avoiding undue influence), which is associated with the right of refusal and withdrawal (without any detriment). Almost all of these elements were described in *Publication 62* on biomedical research: 'The subject has the right to accept the risk voluntarily, and has an equal right to refuse to accept', 'By free and informed consent is meant genuine consent, freely given, with a proper understanding of the nature and consequence of what is proposed', also mentioning that 'consent can be withdrawn at any time by the subjects' (ICRP, 1992). In *Publication 84* on pregnancy and medical radiation, informed consent is regarded as 'doctrine' and it is pointed out that 'there are usually five basic elements to informed consent, which includes whether one is competent to act, receive a thorough disclosure, comprehend the disclosure, act voluntarily, and consent to the intervention' (ICRP, 2000). For vulnerable people with diminished competency, under undue influence, and pregnant women, additional protection in terms of consent and strict risk benefit assessment is required (ICRP, 1992, 2000).

(73) The right to know is another important concept related to transparency. It emerged in the USA in the 1970s in connection with the efforts of the Federal Occupational Safety and Health Administration to ensure that workers benefit from safe and healthy working environments. It has evolved to be defined by the Commission as a requirement to disclose full information on hazardous materials disposed, emitted, produced, stored, used, or simply present in working places or in the environment of communities (e.g. radon, naturally occurring radioactive materials) (ICRP, 2007b, 2014b, 2016).

(74) In publications on environmental protection (ICRP, 2003, 2014a), transparency, which enables social control and vigilance of the public, is also emphasised: 'The principle of informed consent, which emphasises the need for communication and public involvement, starts at the planning stage and well before decisions are taken from which there is no return. Such transparency of decision making should enable analysis and understanding of all stakeholders' arguments, although decisions against certain stakeholders may not be avoided. Transparency is usually secured by way of an environmental impact assessment' (ICRP, 2003).

(75) Finally, accountability and transparency can be mutually reinforcing. Together they allow stakeholders to be aware of up-to-date information required to make informed decisions, and also to possibly participate in the decision-making process.

4.3. Inclusiveness (stakeholder participation)

(76) The value of inclusiveness is usually referred to using the phrase 'stakeholder participation', which is the way the value is applied in practice. Stakeholder

participation, also referred to as 'stakeholder involvement' or 'stakeholder engagement', means 'involving all relevant parties in the decision-making processes related to radiological protection' (IRPA, 2008). In recent decades, stakeholder participation has become an essential part of the ethical framework in private and public sector organisations. Thus, inclusiveness is one of the essential procedural values, along with accountability and transparency, needed to make ethical decisions in organisations. Most likely it was Lauriston Taylor who first suggested engaging with stakeholders in radiological protection. In the Sievert Lecture he gave in 1980, one can read: 'Aside from our experienced scientists, trained in radiation protection, where do we look further for our supply of wisdom? Personally, I feel strongly that we must turn to the much larger group of citizens generally, most of whom have to be regarded as well-meaning and sincere, but rarely well-informed about the radiation problems that they have to deal with. Nevertheless, collectively or as individuals, they can be of great value...in developing our total radiation protection philosophy' (Taylor, 1980).

(77) Engaging stakeholders in radiological protection was first implemented in the late 1980s and early 1990s in the context of the management of exposures in areas contaminated by the Chernobyl accident and sites contaminated by past nuclear activities in the USA (IAEA, 2000). In such exposure situations, citizens found themselves confronted with radioactivity impacting their everyday lives, which posed new questions that the system in place at the time had difficulty in answering. This, in turn, led the Commission to replace the process-based approach of using practices and interventions to a situation-based approach with the distinction between existing, planned, and emergency exposure situations (ICRP, 2007a).

(78) Stakeholder participation was first introduced by ICRP in *Publication 82*: 'Many situations of prolonged exposure are integrated into the human habitat and the Commission anticipates that the decision-making process will include the participation of relevant stakeholders rather than radiological protection specialists alone' (ICRP, 1999). It was further elaborated in *Publication 101b*: 'The involvement of stakeholders is a proven means to achieve incorporation of values into the decision-making process, improvement of the substantive quality of decisions, resolution of conflicts among competing interests, building of shared understanding with both workers and the public, and building of trust in institutions' (ICRP, 2006). It became a requisite in *Publication 103* in relation to the principle of optimisation of protection: 'It should also be noted that the Commission mentions, for the first time, the need to account for the views and concerns of stakeholders when optimising protection' (ICRP, 2007a).

(79) Participation of stakeholders in the decision-making process related to radiological protection is an effective way to take into account their concerns and expectations, as well as their knowledge about the issue at stake. It is also a way for both professionals and stakeholders to better understand what is at stake with the exposure situation. This, in turn, enables adoption of more effective, sustainable, and fair protective actions promoting empowerment and autonomy of stakeholders. Participation of stakeholders in the operation and maintenance of medical,

industrial, and nuclear installations has proven to be an effective way to keep occupational exposures as low as reasonably achievable. Experience from the management of the consequences of the Chernobyl accident and, more recently, the Fukushima accident demonstrated that empowerment of affected people helps them to regain confidence, to understand the situation they are confronted with, and finally to make informed decisions and act accordingly. In other words, engaging stakeholders is a way to respect those affected, and in the case of postaccident situations, to help restore their dignity (Lochard, 2004; ICRP, 2015a).

(80) It is the responsibility of experts and authorities to ensure fair support of all groups of exposed people. Fairness in this respect refers to the core values of justice and dignity. The requirement to be treated fairly is a key condition for those desiring to enter into a dialogue with experts and authorities with the objective to promote well-being and self-determination. This dialogue allows all parties to better understand the situation at stake, and helps empower individuals to make informed decisions. The empowerment process relies on the development of 'practical radiological protection culture' among those involved. This last notion was introduced in *Publication 111*, which is devoted to the protection of people living in long-term contaminated areas after a nuclear accident (ICRP, 2009), but applies to all exposure situations. Practical radiological protection culture can be defined as the knowledge and skills enabling each individual to make well-informed choices and behave wisely when directly confronted with radiation. It is a duty of radiological protection professionals to support making these choices providing access to scientific knowledge and expertise in the spirit of the core ethical values that underlie the system of radiological protection (ICRP, 2009).

(81) A recent ICRP publication on protection of the environment gives explicit procedural recommendations for involving stakeholders effectively: 'Guidelines should be established at the beginning to ensure that the process is effective and meaningful for all parties' and that 'Some of these guidelines include, but are not limited to, the following: clear definition of the role of stakeholders at the beginning of the process; agreement on a plan for involvement; provision of a mechanism for documenting and responding to stakeholder involvement; and recognition, by operators and regulators, that stakeholder involvement can be complex and can require additional resources to implement' (ICRP, 2014a).

5. CONCLUSION

(82) The system of radiological protection is based on three pillars: science, ethics, and experience. As far as ethics is concerned, this publication portrays the system as relying on four core ethical values: beneficence/non-maleficence, prudence, justice, and dignity. Beneficence and non-maleficence are directly related to the aim of preventing or reducing harmful effects for humans and the environment. Prudence allows uncertainties concerning these effects to be taken into account. Justice is the way to ensure social equity and fairness in decisions related to protection. Dignity is to consider the respect that one must have for people.

(83) The principle of justification requires that any decision that alters a radiation exposure situation should do more good than harm. This means that by reducing existing exposures or introducing a new radiation source, the achieved benefit to individuals and society should be greater than the associated disadvantages in terms of radiation risk, but also of any other nature. Thus, the justification principle combines the ethical values of beneficence and non-maleficence, but also the ethical value of prudence, as part of the radiation risk is associated with the use of the LNT model.

(84) The principle of optimisation of protection, in turn, requires that all exposures should be kept as low as reasonably achievable, taking into account economic and societal factors, using restrictions on individual exposures to reduce inequities in the distribution of exposures among exposed groups. This is the cornerstone of the system. On the one hand, it is a principle of action that allows the practical implementation of prudence. On the other hand, it allows the introduction of equity and fairness in the distribution of exposures among people exposed, which refers directly to the ethical value of justice. Ultimately, taking into consideration the particular circumstances in which people are exposed as well as their concerns and expectations, the principle of optimisation is to respect people and treat them with dignity.

(85) The principle of limitation requires that all individual exposures do not exceed the dose limits recommended by the Commission. Like the principle of optimisation, it refers directly to the ethical value of prudence, but more so to justice by restricting the risk in an equitable manner for a given planned exposure situation and category of exposure.

(86) The application of the three principles will depend on the exposure situations and the category of exposure, particularly in medical exposure. Dose limits, for example, do not apply to medical exposures because the balance of the risk and the benefit is specific to the patient in order to provide the best 'margin of benefit over harm'. However, equity is also part of the medical practice through the use of diagnostic reference levels aiming to reduce the frequency of unjustified high or low exposure for a specified medical imaging task. In reality, the ethical considerations are more complex as there is also potential for benefit and harm to others, most notably to the medical staff who also receive some dose, and others such as family and friends who may receive some dose depending on the type of procedure, and who might also gain an indirect benefit derived from the medical benefit to the patient.

(87) Integrated into the three overarching principles of justification, optimisation, and limitation, the core ethical values allow people to act virtuously while taking into account the uncertainties associated with the effects of low dose, and to evaluate the criteria for judging the adequacy of these actions. In practice, the search for reasonable levels of protection (the principle of optimisation) and tolerable exposure levels (the principle of limitation) is a permanent quest that depends on the prevailing circumstances in order to act wisely – in other words, with the desire to do more good than harm (beneficence/non-maleficence), to avoid unnecessary exposure (prudence), to seek fair distribution of exposure (justice), and to treat people with respect (dignity).

(88) The system of radiological protection has also integrated procedural values, particularly accountability, transparency, and inclusiveness, reflecting the importance of responsible behaviour of those involved in the radiological protection process, proper information, and also preservation of the autonomy and dignity of the individuals potentially or actually exposed to radiation.

(89) Until now, the basic aim of the system of radiological protection for humans was to prevent deterministic effects and to keep stochastic effects as low as reasonably achievable, taking into account economic and societal considerations. Recent developments have suggested the inclusion of further aspects of individual and collective well-being of exposed people, such as psychosocial and mental health. This is particularly the case for the management of postaccident situations, as stated in *Publication 111* (ICRP, 2009, Para. 23), with the objective to improve the daily life of exposed individuals.

(90) The inclusion of natural or man-made radiation in existing exposure situations in the latest recommendations of the Commission also highlighted the need to foster the development of an appropriate radiological protection culture within society, enabling each citizen to make well-informed choices and behave wisely in situations involving potential or actual exposure to ionising radiation.

(91) Furthermore, the Commission is also concerned with protection of the environment. Starting with *Publication 91* (ICRP, 2003), a framework has been developed within which the environment can be considered. The Commission considers now that a holistic and integrated view of all the benefits and impacts that may result from the introduction of a new source in planned exposure situations, or consideration of actions in existing and emergency exposure situations, should include appropriate consideration of protection of both people and the environment.

(92) The responsibility of the Commission is to develop the radiological protection system for the public benefit using scientific knowledge and expert judgements. Nevertheless, the Commission believes that by eliciting and diffusing the ethical values and related principles that underpin the radiological protection system, both experts and the public will undoubtedly gain a clearer view of the societal implications of its recommendations. Just as with science and experience, ethics alone is unable to provide a definitive solution to the questions and dilemmas generated by the use or presence of radiation. However, ethics can certainly provide useful insights on the principles and philosophy of radiological protection, and thus help the dialogue between radiological protection professionals and other stakeholders.

REFERENCES

Akabayashi, A., Hayashi, Y., 2014. Informed consent revisited: a global perspective. In: Akabayashi, A. (Ed.), The Future of Bioethics: International Dialogues. Oxford University Press, Oxford, pp. 735–749.

Appiah, K.A., 2006. Cosmopolitanism: Ethics in a World of Strangers. W.W. Norton, New York.

Beauchamp, T.L., Childress, J.F., 1979. Principles of Biomedical Ethics. Oxford University Press, Oxford.

Beauchamp, T.L., Childress, J.F., 1994. Principles of Biomedical Ethics, fourth ed. Oxford University Press, Oxford.

Beauchamp, T.L., Childress, J.F., 2009. Principles of Biomedical Ethics, seventh ed. Oxford University Press, Oxford.

Becquerel, H., 1896. Emission des radiations nouvelles par l'uranium metallique. C. R. Acad. Sci. Paris 122, 1086.

Bok, S., 1995. Common Values. University of Missouri Press, Columbia, MO.

Clarke, R.H., 2003. Changing philosophy in ICRP: the evolution of protection ethics and principles. Int. J. Low Radiat. 1, 39–49.

Clarke, R.H., Valentin, J., 2009. The history of ICRP and the evolution of its policies. ICRP Publication 109. Ann. ICRP 39(1), pp. 75–105.

Clement, C., Lochard, J., 2017. Recent reflections on the ethical basis of the system of radiological protection. In: Zölzer, F., Meskens, G. (Eds.), Ethics of Environmental Health. Routledge, Abingdon, Oxfordshire, pp. 76–85.

Curie, M., 1898. Rayons émis par les composes de l'uranium et du thorium. C. R. Acad. Sci. Paris 126, 1101.

DHEW, 1979. National Commission for the Protection of Human Subjects of Biomedical and Behavioral Research. The Belmont Report. DHEW Publication No. OS 78-0012. Department of Health, Education and Welfare, Washington, DC. Available at: http://videocast.nih.gov/pdf/ohrp_belmont_report.pdf (last accessed 29 November 2017).

Edelstein, L., 1943. The Hippocratic Oath: Text, Translation, and Interpretation. Johns Hopkins University Press, Baltimore, MA.

Forsberg, E-M., 2004. The ethical matrix – a tool for ethical assessment of biotechnology. Glob. Bioeth. 17, 167–172.

Frankena, W.K., 1963. Ethics. Prentice Hall, Englewood Cliffs, NJ.

Fuchs, W.C., 1896. Effect of the Röntgen rays on the skin. West. Electr. December 1896, 291.

Gonzalez, A., 2011. The Argentine approach to radiation safety: its ethical basis. Sci. Technol. Nucl. Install. 2011, 910718.

Habermas, J., 1992. Between Facts and Norms: Contributions to a Discourse Theory of Law and Democracy. MIT Press, Cambridge, MA.

Habermas, J., 1998. The Postnational Constellation. MIT Press, Cambridge, MA.

Hansson, S.O., 2007. Ethics and radiation protection. J. Radiol. Prot. 27, 147–156.

IAEA, 2000. Restoration of Environments with Radioactive Residues. Proceedings of an International Symposium, 29 November–3 December, 1999, Arlington, VA, USA. International Atomic Energy Agency, Vienna, pp. 671–772.

IAEA, 2014. Radiation Protection and Safety of Radiation Sources: International Basic Safety Standards, General Safety Requirements Part 3. IAEA Safety Standards Series No. GSR Part 3. International Atomic Energy Agency, Vienna.

ICRP, 1951. International recommendations on radiological protection. Revised by the International Commission on Radiological Protection and the 6th International Congress of Radiology, London, 1950. Br. J. Radiol. 24, 46–53.

ICRP, 1955. Recommendations of the International Commission on Radiological Protection. Br. J. Radiol. (Suppl. 6), 100.

ICRP, 1959. Recommendations of the International Commission on Radiological Protection. ICRP Publication 1. Pergamon Press, Oxford.

ICRP, 1966. Recommendations of the International Commission on Radiological Protection. ICRP Publication 9. Pergamon Press, Oxford.

ICRP, 1977. Recommendations of the International Commission on Radiological Protection. ICRP Publication 26. Ann. ICRP 1(3).

ICRP, 1983. Cost–benefit analysis in the optimization of radiation protection. ICRP Publication 37. Ann. ICRP 10(2/3).

ICRP, 1991. 1990 Recommendations of the International Commission on Radiological Protection. ICRP Publication 60. Ann ICRP 21(1–3).

ICRP, 1992. Radiological protection in biomedical research. ICRP Publication 62. Ann. ICRP 22(3).

ICRP, 1997a. General principles for the radiation protection of workers. ICRP Publication 75. Ann. ICRP 27(1).

ICRP, 1997b. Radiological protection policy for the disposal of radioactive waste. ICRP Publication 77. Ann. ICRP 27(S).

ICRP, 1998. Radiation protection recommendations as applied to the disposal of long-lived solid radioactive waste. ICRP Publication 81. Ann. ICRP 28(4).

ICRP, 1999. Protection of the public in situations of prolonged radiation exposure. ICRP Publication 82. Ann. ICRP 29(1/2).

ICRP, 2000. Pregnancy and medical radiation. ICRP Publication 84. Ann. ICRP 30(1).

ICRP, 2003. A framework for assessing the impact of ionising radiation on non-human species. ICRP Publication 91. Ann. ICRP 33(3).

ICRP, 2006. The optimisation of radiological protection – broadening the process. ICRP Publication 101b. Ann. ICRP 36(3).

ICRP, 2007a. The 2007 recommendations of the International Commission on Radiological Protection. ICRP Publication 103. Ann. ICRP 37(2–4).

ICRP, 2007b. Radiological protection in medicine. ICRP Publication 105. Ann. ICRP 37(6).

ICRP, 2008. Environmental protection: the concept and use of reference animals and plants. ICRP Publication 108. Ann. ICRP 38(4–6).

ICRP, 2009. Application of the Commission's recommendations to the protection of people living in long-term contaminated areas after a nuclear accident or a radiation emergency. ICRP Publication 111. Ann. ICRP 39(3).

ICRP, 2013. Radiological protection in geological disposal of long-lived solid radioactive waste. ICRP Publication 122. Ann. ICRP 42(3).

ICRP, 2014a. Protection of the environment under different exposure situations. ICRP Publication 124. Ann. ICRP 43(1).

ICRP, 2014b. Radiological protection in security screening. ICRP Publication 125. Ann. ICRP 43(2).

ICRP, 2014c. Radiological protection against radon exposure. ICRP Publication 126. Ann. ICRP 43(3).

ICRP, 2015a. ICRP and Fukushima. ICRP Dialogue Initiative. International Commission on Radiological Protection, Ottawa. Available at: http://new.icrp.org/page.asp?id=189 (last accessed 25 August 2015).

ICRP, 2015b. ICRP Code of Ethics. International Commission on Radiological Protection, Ottawa. Available at: http://www.icrp.org/docs/ICRP%20Code%20of%20Ethics.pdf (last accessed 25 August 2015).

ICRP, 2016. Radiological protection from cosmic radiation in aviation. ICRP Publication 132. Ann. ICRP 45(1).

IRPA, 2004. IRPA Code of Ethics. International Radiation Protection Association, Ottawa. Available at: http://www.irpa.net/members/IRPA%20Code%20of%20Ethics.pdf (last accessed 29 November 2017).

IRPA, 2008. IRPA Guiding Principles for Radiation Protection Professionals on Stakeholder Engagement. International Radiation Protection Association, Ottawa. Available at: http://www.irpa.net/page.asp?id=54494 (last accessed 28 January 2016).

ISO, 2010. Guidance on Social Responsibility. ISO 26000:2010(E). International Organization for Standardization, Geneva.

IXRPC, 1928. International recommendations for x-ray and radium protection. Br. J. Radiol. 1, 359–363.

IXRPC, 1934. International recommendations for x-ray and radium protection. Revised by the International X-ray and Radium Protection Commission and adopted by the 4th International Congress of Radiology, Zürich, July 1934. Br. J. Radiol. 7, 1–5.

Kant, I., 1785. Groundwork of the Metaphysic of Morals [German: Grundlegung zur Metaphysik der Sitten; 1785], translated by H.J. Paton as The Moral Law. Hutcheson, London, 1953, p. 430 (Prussian Academy pagination).

Kimura, R., 2014. Japan, bioethics. In: Jennings, B. (Ed.), Bioethics, Vol. 4, fourth ed. Macmillan Reference, Farmington Hills, MI, pp. 1757–1766.

Küng, H., Kuschel, K-J. (Eds.), 1993. A Global Ethic. The Declaration of the Parliament of the World's Religions. SCM Press, London/Continuum, New York.

Kurihara, C., Cho, K., Toohey, R.E., 2016. Core ethical values of radiological protection applied to Fukushima case: reflecting common morality and cultural diversities. J. Radiol. Prot. 36, 991–1003.

Lapp, R.E., 1958. The Voyage of the Lucky Dragon. Harper & Bros., New York.

Lindell, B., 2001. Logic and ethics in radiation protection. J. Radiol. Prot. 21, 377–380.

Lochard, J., Schieber, C., 2000. The evolution of radiological risk management: an overview. J. Radiol. Prot. 20, 101–110.

Lochard, J., Arranz, L., Gallego, E., Sugier, A., 2004. Living in contaminated territories: a lesson in stakeholder involvement. In: Métivier, H., et al (Eds.), Current Trends in Radiation Protection. EDP Sciences, Les Ulis, pp. 211–220.

Lochard, J., 2013. Stakeholder engagement in regaining decent living conditions after Chernobyl. In: Oughton, D., Hansson, S.O. (Eds.), Social and Ethical Aspects of Radiation Risk Management. Elsevier Science, Oxford, pp. 311–332.

Lochard, J., 2016. First Thomas S. Tenforde topical lecture: the ethics of radiological protection. Health Phys. 110, 201–210.

Malone, J., 2013. Ethical issues in clinical radiology in social and ethical aspects of radiation risk management. Radioact. Environ. 19, 107–129.

Martinez, N., Wueste, D., 2016. Balancing theory and practicality: engaging non-ethicists in ethical decision making related to radiological protection. J. Radiol. Prot. 36, 832–841.

Moody, M., 2011. A Hippocratic Oath for philanthropists. In: Forsyth, D.R., Hoyt, C.L. (Eds.), For the Greater Good of All. Perspectives on Individualism, Society, and Leadership. Palgrave Macmillan, New York, pp. 143–165.

NEA/OECD, 1995. The Environmental and Ethical Basis of the Geological Disposal of Long-lived Radioactive Waste. OECD, Paris.

Nussbaum, M., 2004. Beyond the social contract: capabilities and global justice. Oxford Dev. Stud. 32, 3–16.

Oughton, D., 1996. Ethical values in radiological protection. Radiat. Prot. Dosim. 28, 203–208.

Oughton, D., 2003. Protection of the environment from ionizing radiation: ethical issues. J. Environ. Radioact. 66, 3–18.

Oughton, D., 2008. Public participation – potentials and pitfalls. Energy Environ. 19, 485–496.

Oughton, D., Howard, B., 2012. The social and ethical challenges of radiation risk management. Ethics Policy Environ. 15, 71–76.

Pelligrino, E.D., 2008. Some personal reflections on the 'appearance' of bioethics today. Stud. Bioeth. 1, 52–57.

Roentgen, W.C., 1895. Uber eine neue Art von Strahlen. Sitzungsberichte d. Phys. Mediz. Ges. Wurzburg 9, 132.

Ross, W.D., 1930. The Right and the Good. Oxford University Press, Oxford.

Sandin, P., 2009. Firefighting ethics. Ethic. Perspect. 16, 225–251.

Schneider, T., Lochard. J., Vaillant. L., 2016. The focal role of tolerability and reasonableness in the radiological protection system. Ann. ICRP 45(1S), 322–344.

Schrader-Frechette, K., Persson, L., 1997. Ethical issues in radiation protection. Health Phys. 73, 378–382.

Seedhouse, D.J., 1988. Ethics. The Heart of Health Care. John Wiley, New York.

Sen, A., 2009. The idea of justice. Allen Lane & Harvard University Press, Cambridge.

Silini, G., 1992. Sievert lecture. Ethical issues in radiation protection. Health Phys. 63, 139–148.

Streffer, C., Bolt, C., Follesdal, D., et al., 2004. Low Dose Exposures in the Environment: Dose–effect Relations and Risk Evaluation. Springer Verlag, Berlin.

Streffer, C., Gethmann, C.F., Kamp, G., et al., 2011. Radioactive Waste – Technical and Normative Aspects of its Disposal. Springer-Verlag, Berlin.

Taylor, L., 1957. The philosophy underlying radiation protection. Am. J. Roent. 77, 914–919.

Taylor, L., 1980. Some non-scientific influences on radiation protection standards and practice. The 1980 Sievert lecture. Health Phys. 39, 851–874.

The Interfaith Declaration, 1996. Constructing a code of ethics for international business. In: Business Ethics: a European Review 5, pp. 52–54.

Tsai, D.F.C., 1999. Ancient Chinese medical ethics and the four principles of biomedical ethics. J. Med. Ethics 25, 315–321.

United Nations, 1948. The Universal Declaration of Human Rights. Adopted 10 December 1948. United Nations, New York. Available at: http://www.un.org/Overview/rights.html (last accessed 6 August 2015).

United Nations, 1966. International Covenant on Civil and Political Rights. Adopted and opened for signature, ratification and accession by General Assembly resolution 2200A (XXI) of 16 December 1966, entry into force March 1976, in accordance with Article 49. United Nations, New York.

UNCED, 1992. United Nations Conference on Environment and Development, 3–14 June 1992, Rio de Janeiro, Brazil. Available at: http://www.un.org/geninfo/bp/enviro.html (last accessed 13 August 2015).

UNECE, 2001. Public Participation. The Århus Convention on Access to Information, Public Participation in Decision-making and Access to Justice in Environmental Matters. Adopted June 1998, ratified October 2001. United Nations Economic Commission for Europe, New York. Available at: www.unece.org/env/pp/welcome.html (last accessed 29 November 2017).

UNESCO, 2005. The Precautionary Principle. United Nations Educational, Scientific and Cultural Organization, Paris. Available at: http://unesdoc.unesco.org/images/0013/001395/139578e.pdf (last accessed 29 November 2017).

Valentin, J., 2013. Radiation risk and the ICRP. In: Oughton, D., Hansson, S.O. (Eds.), Social and Ethical Aspects of Radiation Risk Management. Elsevier Science, Oxford, pp. 17–32.

WHO, 1948. Preamble to the Constitution of the World Health Organization as adopted by the International Health Conference, 19 June–22 July 1946, New York, USA. Signed on 22 July 1946 by the representatives of 61 States (Official Records of the World Health Organization, no. 2, p. 100) and entered into force on 7 April 1948. World Health Organization, Geneva.

Zölzer, F., 2013. A cross-cultural approach to radiation ethics. In: Oughton, D., Hansson, S.O. (Eds.), Social and Ethical Aspects of Radiation Risk Management. Elsevier Science, Oxford, pp. 53–70.

Zölzer, F., 2016. Are the core values of the radiological protection system shared across cultures? Ann. ICRP 45(1S), 358–372.

ANNEX A. ETHICAL THEORIES

(A1) This annex provides a brief summary of some of the theories of ethics that have been referred to in exploring the ethical foundation of the radiological protection system. These theories can be characterised as 'Western', from ancient Greek to modern German and British philosophy.

(A2) Understanding the main points of these theories may help to track some of the conflicts or dilemmas that occur in practical radiological protection. Of course, understanding certain theories does not, in itself, provide a solution to an issue, and the Commission has never taken a position of preferring one theory over another. Nonetheless, knowledge of these theories may facilitate mutual understanding among people advancing different arguments.

(A3) Ethics is a discipline of philosophy that discusses virtue and vice (character), good and bad (quality), or right and wrong (action). The terms 'ethics' and 'moral philosophy' are largely used to describe the same exercise. The origin of the former is Greek, and that of the latter is Latin. 'Morals' is sometimes used to describe culturally and religiously based values and norms.

(A4) There are three main levels of ethical theory often referred to in discussions of radiological protection: meta-ethics (discussing the general meaning of ideas such as 'virtue', 'good', or 'right'), normative ethics (discussing how one should act, and which values and norms should be followed), and applied ethics (discussing specific issues, e.g. in medicine or engineering, based on ethical theories or principles).

(A5) Within normative ethics, three main theories can, in turn, be identified that have been used to discuss the radiological protection system. These are: virtue ethics (discussing virtuous life based on a certain concept of human nature); deontological ethics (discussing a set of obligations or rules for human society); and consequentialist ethics (discussing the preferability of certain actions on the basis of their outcomes).

(A6) The ethics of radiological protection has some affinity with other fields of applied ethics, such as biomedical ethics (see Annex B), environmental ethics, engineering ethics, etc. The literature on these topics is quite diverse, but only a few ICRP publications address similar questions with respect to radiation [e.g. *Publications 62* (ICRP, 1992) and *91* (ICRP, 2003)].

(A7) An article on the history of ICRP and the evolution of its policies (Clarke and Valentin, 2009) provides analysis of the recommendations of the Commission from its beginning, and comes to the conclusion that they focused primarily on three theories of ethics: (1) early recommendations (1928–1950s) focusing on virtue ethics; (2) intermediate recommendations (1960s–70s) focusing on utilitarian ethics (the most well-known version of consequentialism); and (3) present recommendations (80s to present) focusing on deontological ethics. The intention of this analysis is to emphasise that in developing its recommendations, the Commission has attempted to make a balance that needs to be reached among these theories.

(A8) The following is a short summary of how the three theories of normative ethics are related to radiological protection.

A.1. Virtue ethics

(A9) Representatives of this theory are the ancient Greek philosophers Plato (BC427–347) and Aristotle (BC384–322). They based their reasoning on the moral nature or characteristics of the human being, rather than on rules or obligations. Good is what a good or virtuous person would do. If one considers deterministic radiation effects for instance, this idea can be simply linked to human nature, which tends to avoid harm. More generally, the 'justification' principle of radiological protection can be understood as expressing the same idea, as it relies on human nature not only avoiding harm but also doing good. In other words, it is the right motivation of a human following his or her moral nature that leads to the right action (Hansson, 2007).

A.2. Consequentialist ethics

(A10) The most well-known version of consequentialism is utilitarianism, and the representatives of this theory are the English scholars Jeremy Bentham (1748–1832) and John Stuart Mill (1806–1873). They maintained that the only valid criterion of the goodness of an act or a rule is its good consequences, rather than the good nature of a human being or obligations in human society. The most well-known notion of utilitarianism is that we should strive for 'the greatest happiness of the greatest number'. The 'optimisation' principle is often linked to this utilitarian approach, as it seeks to keep radiation exposures 'as low as reasonably achievable, taking into account economic and societal factors'. This principle is associated with the risk of stochastic effects, especially at low doses. In the past, it has often been understood to suggest decision making based on cost–benefit analysis to calculate the greatest financial gain for society, while allowing only the smallest sacrifice of individuals. Consequentialist ethics does not always seek to maximise collective gain, but it is sometimes used to balance risk and benefit for an individual.

A.3. Deontological ethics

(A11) A very important representative of this theory is the German philosopher Immanuel Kant (1724–1804). Kant argued that human beings possess a rational nature and have the capacity of self-regulation, which is called 'autonomy'. Good will leads them to act according to their duty, or the moral law. Kant asserted that one should not treat human beings merely as means to an end, but rather as ends in themselves. This means that we should not sacrifice an individual to achieve 'the greatest happiness of the greatest number'. At the same time, it means that we should respect every individual's free choice. Another version of deontological ethics discussed in radiological protection is that developed by the Scottish philosopher William David Ross (1877–1971). He is well known as a translator of Aristotle's works and wrote much about Greek philosophy, so it is not surprising that his theory also includes some elements from virtue ethics. Ross provided a set of prima-facie duties (fidelity, reparation, gratitude, non-maleficence, justice, beneficence, self-improvement) that help determine what a person ought to do, with the proviso that one or the other may take precedence in a particular situation. With regards to the principles of radiological protection, 'limitation' can be linked directly to deontological ethics. This notably applies to the idea that individuals need to be protected in an equitable manner, and therefore limits should be set to avoid sacrificing one person for the sake of others. In addition, 'stakeholder participation' in the decision-making process is based on respecting each person's human dignity. Therefore, the idea that radiological protection today has come to rely more heavily on deontological ethics cannot be denied, although deriving the principles of radiological protection from Western ethical theories still requires referring to virtue ethics and utilitarianism as well. In practice, the different perspectives of all three theories have to be brought to bear.

ANNEX B. BIOMEDICAL ETHICAL PRINCIPLES

(B1) Much of the discussion about the ethics in radiological protection referred to the three theories of normative ethics mentioned in Annex A, but there is also some reference to applied ethics. One of the most widely discussed frameworks in applied ethics is that developed by Beauchamp and Childress (1979) on biomedical ethics. Their initial aim was to find principles that the former, as a utilitarian, and the latter, as a deontologist, could agree to without referring to a particular single theory of ethics. The resultant system is not based on one unique ethical framework, but on four principles:
- respect for autonomy (the norm of allowing individuals to decide for themselves);
- non-maleficence (the norm of avoiding the causation of harm);
- beneficence (a group of norms for providing benefits); and
- justice (a group of norms for distributing benefits, risks, and costs fairly)

(B2) Beauchamp and Childress argued that both the utilitarian and the deontologist could agree fully with all four principles, and would find them ethically and morally relevant, albeit for different reasons. Some discussion may arise when it comes to balancing these principles: deontologists tend to prioritise 'non-maleficence' over 'beneficence', whereas utilitarians would rather carry out a cost–benefit assessment, maximising benefit and minimising harm. The Belmont Report (DHEW, 1979) issued by the US National Commission for the Protection of Human Subjects of Biomedical and Behavioral Research took on a similar style, and suggested three principles of ethics for research involving human subjects: respect for persons (instead of autonomy); beneficence (including non-maleficence as a component); and justice. Beauchamp was one of the main contributors of the Belmont Report.

(B3) These three or four principles have come to be known as the principles of 'bioethics', which emerged in the 1960s to 1970s in the USA. These principles have also been widely adopted in other areas, including public and environmental health ethics (Seedhouse, 1988), technology assessment (Forsberg, 2004), firefighting ethics (Sandin, 2009), and, within radiological protection, as the basis of an ethical evaluation of remediation strategies (Oughton, 2003).

(B4) The framework was not originally conceived as a cross-cultural type of ethics. When Beauchamp and Childress introduced the term, they just claimed that 'all morally serious persons' (Beauchamp and Childress, 1994) or – in a later version – 'all persons committed to morality' (Beauchamp and Childress, 2009) would agree with their four principles. With time, they developed the notion that the principles could be rooted in 'common morality', which is 'not relative to cultures or individuals, because it transcends both' (Beauchamp and Childress, 2009). Attempts have been made to show that the principles of biomedical ethics can indeed be traced in various cultural, religious, and philosophical contexts around the world, in particular in their most respected written and oral traditions (Zölzer, 2013).

(B5) In this context, of course, arguments against commonality and for cultural variety have also been put forward. Reflecting the prominent status which these principles have gained, a number of criticisms have been brought to bear against the 'Georgetown Mantra' (so called because this set of principles was generated at Georgetown University).

(B6) The first type of criticism is that the three or four principles tend to be used somewhat casually for the analysis of complicated issues, without deep deliberation about the situation with which an individual may be confronted. Critics coming from that perspective prefer to consider each case by means of a situation-based or narrative approach, rather than one based on principles.

(B7) Another type of criticism is that although these principles are contained in Western as well as non-Western theories, there are some differences. For example, 'autonomy' emphasises the individual's right of self-determination for Westerners, but many non-Westerners will prefer 'related autonomy' (Kimura, 2014) such as family or community-based decision making (Akabayashi and Hayashi, 2014). Also, 'justice' is largely understood as equity in the West, but in some non-Western cultural contexts, equal rights have not been widely established because of a traditional concern about social hierarchy.

ANNEX C. CROSS-CULTURAL VALUES

C.1. The rise of global ethics

(C1) Global approaches to questions of values and norms may seem to be fraught with difficulties, but the fact is that people around the world are moving closer and closer together, and there is a growing need for common perspectives. A milestone in this development was certainly the Universal Declaration of Human Rights adopted by the United Nations General Assembly in 1948 (United Nations, 1948). This was a vow of the international community never again to allow such atrocities as happened during World War II, caused, in part, because of a lack of shared values and norms among people. It led to two multilateral treaties, the International Covenants on Civil and Political Rights and on Economical, Social, and Cultural Rights (United Nations, 1966). In the second half of the 20th century, and especially around the turn of the 21st century, a number of other international statements on human rights followed, as shown in Table C.1.

(C2) It should be noted that many countries in the world have still not ratified all of the above set of declarations on human rights. There are also some countries which have ratified them, but human rights have not been sufficiently established in reality. To give assurances that these declarations work in concrete situations, it is still necessary to look for universally accepted values and norms with relevance for particular subject areas. Radiological protection is only one of these areas.

(C3) With the rise of globalisation over the last few decades, philosophers have addressed the general need for, and possibility of, global ethics from various points of view. A few examples may suffice here. Habermas speaks of a 'post-national constellation' in which we find ourselves, and claims that 'world citizenship . . . is already taking shape today in worldwide political communications' (Habermas, 1992, 1998). Interested in human flourishing and its global dimension, Sen has written extensively about the 'idea of justice', which he shows to be central to various cultures around the world, past and present (Sen, 2009). One of his close associates, Nussbaum has identified a number of 'core capabilities' to which all individuals in all societies should be entitled, thus constituting the base of her account of 'global justice' (Nussbaum, 2004). Appiah explores the reasonability of cosmopolitanism,

Table C.1. A few milestones in the development of global values and norms.

1948 Universal Declaration of Human Rights
1959 Declaration of the Rights of the Child
1966 International Covenant on Civil and Political Rights
1966 International Covenant on Economical, Social, and Cultural Rights
1972 Declaration on Human Environment
1992 Declaration on Environment and Development (UNESCO)
1997 Universal Declaration on the Human Genome and Human Rights
2005 Universal Declaration on Bioethics and Human Rights

which he defines as 'universality plus difference'. While emphasising 'respect for diversity of culture', he suggests there is 'universal truth, too, though we are less certain that we have it all already' (Appiah, 2006). Finally, Bok suggests that 'certain basic values [are] necessary to collective survival' and therefore constitute a 'minimalist set of such values [which] can be recognised across societal and other boundaries'. That does not preclude the existence of 'maximalist' values, usually more culture specific, nor the possibility that they can 'enrich' the debate, and the 'need to pursue the enquiry about which basic values can be shared across cultural boundaries is urgent' (Bok, 1995).

(C4) One area in which cross-culturally shared ethical principles, values, and norms are actively discussed is interfaith dialogue. An outcome of such activities was the Declaration towards a Global Ethic signed at the Parliament of the World's Religions 1993 in Chicago by the representatives of more than 40 different religious groups. It proceeded from the assumption that 'there already exist ancient guidelines for human behaviour which are found in the teachings of the religions of the world and which are the condition for a sustainable world order' (Küng and Kuschel, 1993). Interfaith declarations on more specific topics such as business ethics and environmental ethics have followed (The Interfaith Declaration, 1996).

C.2. A short review of the core values in different cultural contexts

(C5) In order to validate the assumption that the core values identified as founding the radiological protection system are shared across cultures, one could, of course, think of empirical research, but investigations along these lines have not been undertaken systematically, and their results would just reflect people's current dispositions. Orientation in matters of ethics has been provided throughout the ages by the religious and philosophical traditions of the different cultures, and in spite of a tendency towards secularisation in many societies, they continue to have great influence. It is therefore of interest for the purpose of this publication to look at a few such sources, and assess (to the degree possible within limited space) the universality of the values identified as fundamental for the system of radiological protection. It should be noted that the construction of a set of values which are identified as core values of the radiological protection system does not mean that this set is universally applicable to all aspects of life in all cultures. Each of these values can be found in various cultural contexts, but their weight can certainly vary across cultures and even within one culture depending on what issue is discussed.

C.2.1. Beneficence and non-maleficence

(C6) 'To abstain from doing harm' is one of the central features of the Hippocratic Oath (Edelstein, 1943), which was later adopted by Jewish, Christian, and Muslim physicians (Pelligrino, 2008). The principle is also mentioned, albeit indirectly, in similar texts from ancient China (Tsai, 1999). Of course, it has always been understood that sometimes pain has to be inflicted to achieve healing, and thus non-maleficence has to be balanced with beneficence. To work 'for the good of the patient' is also part of the Hippocratic Oath, and it features quite prominently in the mentioned Chinese medical texts.

(C7) More generally (i.e. outside the context of medicine), both beneficence and non-maleficence can be seen as core principles in any system of religious ethics. A central concept of both Hinduism and Buddhism is 'ahimsa' which means kindness and non-violence to all living beings. Both the Torah and the Gospel express the same thought in a different way by exhorting everybody to 'love your neighbour as yourself', and Islamic jurisprudence has the guideline that 'if a less substantial instance of harm and an outweighing benefit are in conflict, the harm is forgiven for the sake of the benefit'.

(C8) When it comes to 'taking into account economic and societal factors', as stipulated by the principle of optimisation, the interest of the general public, the 'common good', is a related concept of importance that is also shared across cultures. All religious writings exhort their readers to solidarity with the underprivileged in society, as is, for instance, expressed in one of the Psalms, 'Blessed is the one who is considerate of the destitute'.

(C9) More generally, the traditions remind us that we are not just individuals. An African proverb says 'A single tree cannot make a forest', and highlights that African ethics privileges the common good and a sense of duty to the public over personal or individualistic motives. In the play 'Muntu', Joe de Graaft demonstrates that the individual's needs – peace, freedom, dignity, and security – can only be protected and guaranteed by the community. John Mbiti asserts, 'I am, because we are; and since we are, therefore I am'.

C.2.2. Prudence

(C10) In recent decades, there have been numerous talks about the 'precautionary principle', especially in the context of environmental issues. Of course, the principle in its modern form cannot be expected to appear in the written and oral traditions of different cultures. Exhortations to prudence, however, are ubiquitous, and they are generally interpreted, by people referring to those traditions for orientation, as suggesting a precautionary approach.

(C11) A Hindu text suggests to 'act like a person in fear before the cause of fear actually presents itself', whereas Confucius simply says that 'The cautious seldom err'. In the biblical Proverbs, one finds the following statement: 'Those who are prudent see danger and take refuge, but the naïve continue on and suffer the consequences'. A representative of the Australian Aboriginals and Torres Strait Islanders has stated: 'Over the past 60,000 years we, the indigenous people of the world, have successfully managed our natural environment to provide for our cultural and physical needs. We have no need to study the non-indigenous concepts of the precautionary principle [and others]. For us, they are already incorporated within our traditions'.

C.2.3. Justice

(C12) The 'Golden Rule', the first principle of justice and altruism, claims 'Do unto others what you want them to do unto you', and is one of the most common ethical guidelines around the world. It is found in every single tradition one may choose to look at, and even its wording is strikingly uniform. A few examples must suffice: 'Hurt not others in ways that you yourself would find hurtful' (Buddhist); 'Therefore whatever you want people to do for you, do the same for them, because this summarises the Law and the Prophets' (Christian); 'If thine eyes be turned towards justice, choose thou for thy neighbour that which thou choosest for thyself' (Bahá'í).

(C13) In African ethics, this principle has ontological, religious, and communal implications. The main basis of the principle is the concept of empathy. Empathy helps a person to imagine the effects of an action or of the failure to act on oneself before considering what it would mean for others, and thus is conducive to 'cooperation, solidarity and fellowship'.

(C14) Justice, as such, is verifiably an element of common morality. The Bhagavad Gita contains the promise that 'He who is equal-minded among friends, companions and foes... among saints and sinners, he excels'. The Psalms observe that, 'He loves righteousness and justice; the world is filled with the gracious love of the Lord', whereas Muhammad advises his followers to be 'ever steadfast in upholding equity..., even though it be against your own selves or your parents and kinsfolk'.

(C15) A look at secular philosophy will also be instructive here, as justice has been of prime importance since antiquity. Aristotle, for instance, distinguished between different forms of justice, and his analysis has exerted decisive influence on later thought. His concept of 'distributive justice' concerns the allocation of goods and burdens, of rights and duties in a society. About this, he states, 'The only stable state is the one in which all men are equal before the law'.

C.2.4. Dignity

(C16) This last core value is expressed in different ways around the world, but the basic idea is virtually ubiquitous – that of a dignity pertaining equally to all humans. In the Bhagavad Gita, one finds, 'I am the same to all beings... In a Brahma... and an outcast, the wise see the same thing'. In the Bible, the prophet Malachi asks, 'Do we not have one father? Has not one God created us?', and in the Quran, it is expressed as, 'We have conferred dignity on the children of Adam... and favoured them far above most of Our creation'.

(C17) These are just short glimpses from different religious sources, but the broad agreement on the notion that all human beings share the same dignity is also reflected in the Declaration Toward a Global Ethic of the Parliament of World's Religions in 1993. It says that 'every human being without distinction of age, sex, race, skin colour, physical or mental ability, language, religion, political view, or national or social origin possesses an inalienable and untouchable dignity, and everyone, the individual as well as the state, is therefore obliged to honour this dignity and protect it' (Küng and Kuschel, 1993).

(C18) Moreover, human dignity has, for centuries, been invoked by secular philosophers. This strand of thought begins with Stoicism, continues through the Renaissance, and leads up to Enlightenment. In our time, together with the above-mentioned religious traditions, it has played a very prominent role in the drawing up of the Universal Declaration of Human Rights of 1948, and the Universal Declaration of Bioethics and Human Rights of 2005, as mentioned at the beginning of this annex.

C.3. Confucian theory and Asian perspectives

(C19) It would certainly be interesting to discuss the ethics of different cultures one by one, understand their internal logic, and then relate them to the ICRP system of radiological protection. As there is no space to do that here, it was decided to have a closer look at just one system of non-Western ethics, namely Confucian theory, because there has been some discussion over the last decades about 'Asian perspectives', and even 'Asian values', which were allegedly different from those 'forced upon the world' by the West.

(C20) In spite of such claims about fundamental differences between Western and non-Western moral philosophies, Confucian thought in everyday life emphasises moral values that are quite compatible with Western ideas. The fundamental standpoint of Confucianism is that all humans have a disposition towards the good, and are naturally inclined to follow the virtuous model. The five moral values (or virtues) that are embedded in Confucianism are as follows: Ren (仁, benevolence), Yi (義, justice), Li (禮, courtesy), Zhi (智, wisdom), and Xin (信, trust).

(C21) 'Ren' is the foremost value, which integrates all the other four and is an obligation of altruism and humaneness towards other individuals. 'Yi' is a tool for the practice of Ren and is the upholding of righteousness and the moral disposition to do good. 'Li' is the traditional and customary norm, which determines how a person should act properly in everyday life, especially when relating to others. 'Zhi' is the mental ability to understand quickly and correctly the principle of the matter, and to make a right and fair decision. 'Xin' is the trust that should be built among people.

(C22) Fig. C.1 shows the relationship of the five Confucian ethical values related to Western or globally accepted ethical values and principles embedded in the ICRP system of radiological protection.

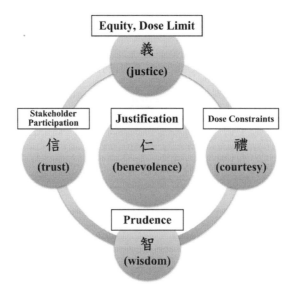

Fig. C.1. Core value system of classic Confucianism connected to core values of the radiological protection system (Kurihara et al., 2016).

(C23) It is obvious that 'Ren' (benevolence) is almost the same concept as beneficence, the former describing rather a disposition, the latter describing a way of acting. Both are widely accepted not only in Western but also in Asian cultural contexts. In Confucian theory, it is often argued that 'Ren' is stronger than other values, and this can give rise to a paternalistic understanding of the value system. Meanwhile, as mentioned above (see Annex B), there is an international consensus not to presuppose a fixed hierarchy between beneficence, non-maleficence, justice, and autonomy (or human dignity).

(C24) 'Yi' (justice) is mostly the same as justice in the Western context. However, in Confucian theory, it also has the meaning of 'royalty' and implies respect for the hierarchy in the society, rather than equal rights of individuals.

(C25) 'Li' (courtesy) means respect for the dignity of a person; however, it is not usually understood to be directly connected to basic human rights of self-determination and equality. It rather implies respect for elders or persons of a higher position within the hierarchy, as well as respect for traditional customs or regulations rather than the individual's freedom.

(C26) 'Zhi' (wisdom) is related to 'prudence', but has a wider meaning. It encompasses the integration of various conflicting values.

(C27) 'Xin' (trust) is achieved through honesty, sincerity, and good faith, and is thus closely related to the modern concepts of accountability, transparency, and stakeholder participation.

(C28) As described here, the implications of 'benevolence/beneficence' and 'wisdom/prudence' are almost the same in Western and Confucian thinking, whereas 'dignity' and 'justice' as the basis of fundamental human rights and equality have been developed in the Western world and the consensus reached there is not necessarily shared by people with a Confucian background, although certain fundamental aspects of the two concepts are universal.

ANNEX D. PARTICIPANTS AT THE WORKSHOPS ON THE ETHICS OF THE SYSTEM OF RADIOLOGICAL PROTECTION

1st Asian Workshop on the Ethical Dimensions of the Radiological Protection System, 27–28 August 2013, Daejeon, Korea
Organised by the Korean Association for Radiation Protection
Hosted by the Korea Institute of Nuclear Safety

Min Baek	Chan Hyeong Kim	Seong-Ho Na
Marie-Claire Cantone	Il-Han Kim	Viet Phuong Nguyen
Kun-Woo Cho	Jong Kyung Kim	Enkhbat Norov
Hosin Choi	Kyo-Youn Kim	Hiroko Yoshida Ohuchi
Mi-Sun Chung	Sung Hwan Kim	Woo-Yoon Park
Christopher Clement	Chieko Kurihara-Saio	Ronald Piquero
Moon-Hee Han	Dong-Myung Lee	Sang-Duk Sa
Sungook Hong	Hee-Seock Lee	Sohail Sabir
Seoung-Young Jeong	JaiKi Lee	John Takala
Kyu-Hwan Jung	Senlin Liu	Man-Sung Yim
Keon Kang	Jacques Lochard	Song-Jae Yoo

1st European Workshop on Ethical Dimensions of the Radiological Protection System, 16–18 December 2013, Milan, Italy
Organised by the Italian Radiation Protection Association and the French Society for Radiological Protection

Marie Barnes	Eduardo Gallego	Guido Pedroli
François Bochud	Alfred Hefner	Francois Rollinger
Giovanni Boniolo	Dariusz Kluszczynski	Thierry Schneider
Marie-Charlotte Bouesseau	Chieko Kurihara-Saio	Michael Siemann
Marie-Claire Cantone	Ted Lazo	John Takala
Kun-Woo Cho	Jean-François Lecomte	Richard Toohey
Christopher Clement	Bernard Le Guen	Emilie van Deventer
Roger Coates	Jacques Lochard	Sidika Wambani
Renate Czarwinski	Jim Malone	Dorota Wroblewska
Daniela De Bartolo	Gaston Meskens	Margherita Zito

Biagio Di Dino   Celso Osimani   Friedo Zölzer
Marie-Helene El Jammal   Deborah Oughton

1st North American Workshop on Ethical Dimensions of the Radiological Protection System, 17–18 July 2014, Baltimore, MD, USA

Organised by the US Health Physics Society, Canadian Radiation Protection Association, and the Mexican Society for Radiological Protection

Ralph Anderson   Nobuyuki Hamada   Yasuhito Sasaki
Edgar Bailey   Raymond Johnson   Glenn Sturchio
Mike Boyd   Ken Kase   Richard Toohey
Dan Burnfield   Toshiso Kosako   Brant Ulsh
Donald Cool   Cheiko Kurihara-Saio   Richard Vetter
Renate Czarwinski   Ted Lazo   Harry Winsor
Yuki Fujimichi   Jacques Lochard

2nd European Workshop on Ethical Dimensions of the Radiological Protection System, 4–6 February 2015, Madrid, Spain

Organised by the Spanish Society for Radiological Protection, Italian Society for Radiological Protection, French Society for Radiological Protection, and UK Society for Radiological Protection

Antonio Almicar   Eduardo Gallego   María Pérez
Marie Barnes   Cesare Gori   Volha Piotukh
François Bochud   Klazien Huitema   Thierry Schneider
Francesco Bonacci   Dariusz Kluszczynski   Patrick Smeesters
Marie-Charlotte Bouesseau Chieko Kurihara-Saio   Behnam Taebi
Marie-Claire Cantone   Jean François Lecomte   John Takala
Pedro Carboneras   Bernard Le-Guen   Jim Thurston
Kun-Woo Cho   Jacques Lochard   Richard Toohey
Christopher Clement   Jim Malone   Eliseo Vaño
Roger Coates   Gaston Meskens   Dorota Wroblewska
Marie-Helène El Jammal   Mohamed Omar   Friedo Zölzer
Sebastien Farin   Deborah Oughton

2ⁿᵈ North American Workshop on Ethical Dimensions of the System of Radiological Protection, 10–12 March 2015, Cambridge, MA, USA

Organised by the Harvard Kennedy School, Belfer Center, Harvard University, and ICRP

Kun-Woo Cho
Christopher Clement
Andrew Einstein
Stephen Gardiner
Nobuyuki Hamada

Bjørn Morten Hofmann
Sheila Jasanoff
Cheiko Kurihara-Saio
Jacques Lochard
Nicole Martinez

Gina Palmer
Laura Reed
Behnam Taebi
John Takala
Friedo Zölzer

2ⁿᵈ Asian Workshop on the Ethical Dimensions of the System of Radiological Protection, 2–3 June 2015, Fukushima, Japan

Organised by Fukushima Medical University and ICRP

Tazuko Arai
Kathleen Araujo
Ryoko Ando
Cécile Asanuma-Brice
Marie-Claire Cantone
Christopher Clement
Aya Goto
Nobuyuki Hamada
Toshimitsu Homma
Audrie Ismail
Wataru Iwata
Michiaki Kai
Mushakoji Kinhide

Mariko Komatsu
Atsushi Kumagai
Chieko Kurihara-Saio
Ted Lazo
Jean-François Lecomte
Jacques Lochard
Nicole Martinez
Hideyuki Matsui
Gaston Meskens
Michio Miyasaka
Makoto Miyazaki
Toshitaka Nakamura
Ohtsura Niwa

Sae Ochi
Deborah Oughton
François Rollinger
Kiriko Sakata
Hisako Sakiyama
Yasuhito Sasaki
Thierry Schneider
Lavrans Skuterud
Megumi Sugimoto
John Takala
Toshihide Tsuda
Fumie Yamaguchi